Popular Annuals of Eastern North America 1865–1914

GARDENING.

POPULAR ANNUALS

OF
EASTERN NORTH AMERICA
1865–1914

PEGGY CORNETT NEWCOMB

DUMBARTON OAKS
Research Library and Collection
Washington, D.C.

Copyright © 1985 by Dumbarton Oaks
Trustees for Harvard University, Washington, D.C.

Frontispiece:

"Gardening." *Washburn and Co's Amateur Cultivator's Guide to the Flower and Kitchen Garden*, Boston, 1869.

Library of Congress Cataloging in Publication Data

Newcomb, Peggy Cornett.
  Popular annuals of eastern North America, 1865–1914.

  Revision of thesis (M.S.)—University of Delaware, 1981.
  Bibliography: p.
  Includes index.
  1. Annuals (Plants)—United States—History.
  2. Annuals (Plants) industry—United States—History.
  I. Title.
  SB422.N48 1984   635.9′312′0974   84-1674
  ISBN 0-88402-138-6

Composition by Graphic Composition, Inc.

Printed by Meriden-Stinehour Press

Contents

Preface   vii

Introduction   1

I   Seed Production and Distribution during an Era of Change   5
       Early Development of the American Seed Industry   5
       The Industrial Revolution: Growth and Change in the Victorian Age   7

II   The Role of Cultivation, Commerce, and Taste in the Development of Annuals   13
       Defining Annuals   13
       Annuals Acquire New Significance in the Trade   17
       Popularizing the Nomenclature   20
       The Role of Annuals in Successive Garden Styles   21

III   History and Development of a Selection of Annuals through the Mid-Victorian Era   35
       Old-Time Favorites and Early Nineteenth-Century Introductions from Abroad   36
       Early Nineteenth-Century Introductions Native to the United States   57

IV   The Impact of Public Events   67
       Major Exhibitions: The 1876 Centennial and the 1893 World's Columbian Exposition   68
       Flower Shows and Trials   73

V   Development of Annuals from the 1890s to 1914   77

| VI | Some General Observations | 105 |

Appendix I   Four Pre-1865 Lists of Annuals   107

Appendix II   A Chronological Documentation of Annuals through the Trade   145

Notes   183

Selected Sources of Information   191

Index   199

Note on the Illustrations

Illustrations of plants in the text are taken from American seed catalogues of the period under discussion.

NEW YELLOW PHLOX, ISABELLINA.

Preface

THIS BOOK IS A REVISION of a manuscript which I submitted to the faculty of the University of Delaware in December 1981 in partial fulfillment of the Longwood Program for the degree of Master of Science in Ornamental Horticulture. My interest in the historical aspects of gardens and plants started with my involvement in restoring the eighteenth- and early-nineteenth-century gardens of Old Salem, a Moravian village restoration in Winston-Salem, North Carolina, and continued during my employment as the resident gardener of a private estate designed by Ellen Biddle Shipman in the 1920s. During the fall of 1979 when, as a first year Longwood Fellow, I was concentrating on a thesis topic, I quite naturally focused upon an area of research which could perhaps address some of the problems encountered in attempts to accurately document and represent authentic plants in restored or recreated gardens. Because I found so much research was already directed toward unraveling the evolution of garden vegetables, I chose instead to narrow my investigation to the changes in annual flowers. Ms. Julia F. Davis, noted garden historian and lecturer, played a key role in helping me to define the historical and geographical parameters which provided the basis for my research. Her broad knowledge of the role of plants in the landscapes of the past century proved invaluable.

During my research I had the privilege of visiting a number of respected horticultural libraries, both public and private, and spending many enjoyable hours exploring their rare book and catalogue collections. To Ms. Dolores Altemus and Mr. Stuart Dick, University of Delaware Library; Ms. Enola Teeter, Longwood

Gardens Library; Ms. Mary Lou Wolf, Pennsylvania Horticultural Society Library; Ms. Elisabeth Woodburn, Booknoll Farm; Ms. Laura T. E. Byers, Dumbarton Oaks Library; and Mr. Charles van Ravenswaay, I extend special thanks for their generous assistance.

The original editing of my manuscript as a thesis was provided by Harold Bruce, author and presently Curator of Collections at Winterthur Museum and Gardens in Wilmington, Delaware. Elisabeth B. MacDougall, Director of Studies in the History of Landscape Architecture, and Lois Fern, Editorial Associate at Dumbarton Oaks, are to be commended for their careful reshaping of my manuscript into a book which, it is hoped, will prove an informative tool for the garden historian and layman alike.

I am particularly indebted to Dr. Richard W. Lighty, Director of Mt. Cuba Center and former coordinator of the Longwood Program, who has supported this project from its initial conception, through its numerous drafts, to its final publication. For his genuine interest and continued assistance I am especially grateful.

Introduction

IN EXAMINING SITES that were the gardens of a century or more ago, landscape historians look to the physical remnants for indications of the shapes of the past. But annuals, the most ephemeral of ornamental plants, leave no impressions on the historic landscape—no discolored imprints of their decayed roots in the soil, no vegetative regenerations from woody crowns. If escaped flowers persist, reseeding themselves yearly among the weeds, it can be argued that they are descendants that prove their ancestors once existed in cultivation. But these echoes of the past fix neither time nor place. Generations of reversion have most likely taken the flower's present aspect back toward its aboriginal state, its state before the forms were developed that the late-nineteenth-century gardener knew. The dilemma remains. Among the tangible artifacts yielding to the archaeologist's trowel, annuals will remain the most elusive of garden elements.

To write then of annuals over the fifty-year period following the Civil War requires investigation away from the site and into the written documents, lists, catalogues, and illustrations that depicted them. Yet there we find that the flowers cannot escape the distortions and opinions of another age. It is impossible to lift them from the pages in a pure state. The basic theme of this study, therefore, is the consideration of a small sample of the many annuals available during the late nineteenth century as they were known and liked by the average American—as they were made "popular."

Annuals prevalent in late-nineteenth-century gardens fall naturally into two groups. There were those already "old-time" favorites from the eighteenth century—balsam, China aster, cockscomb, marigold, mignonette, and nasturtium—which were highly selected, refined, and commonly associated with a variety of classic gardening styles throughout the nineteenth century. The pansy and sweet pea are special cases in this group. Although both had existed demurely in the corners of gardens for some time, they were not seriously considered as ornamentals until vastly improved during the 1800s.

A second category includes flowers introduced into cultivation during the first half of the nineteenth century. These newcomers created a different kind of excitement and enthusiasm. During the latter part of the century, new and unusual forms and colors of petunia, portulaca, and zinnia were being rapidly selected by breeders in continental Europe and in England. Further, the explorations of the western regions of America were revealing wild flowers of great potential for the garden. Flowers such as abronia, California poppy, clarkia, collinsia, Drummond's phlox, gaillardia, and gilia were symbols of discovery, progress, and adventure.

Through the medium of print, all were colorfully depicted in an endless array of new forms, hues, and habits. Within this context, annuals can be regarded as reflections of general cultural impulses, tastes, and trends during an age when "horticultural interest was one of the many side effects of changes in American life and attitudes."[1]

The annuals cited above have been selected as a representative collection rather than as a definitive list of all the flowers available throughout the period. Each will be considered in detail to depict the variety of vogues affecting the development and use of flowers in general. This study should establish the kinds of annuals in use in the second half of the nineteenth century and the evolution in their types that took place in that period. It is hoped that it will be used as a guide by the recreators of historical gardens. It should also draw attention to the need to preserve the fragile and ephemeral documents that were the basis of the research: popular magazines, catalogues, lists, and other transient literature. Being the medium through which new discoveries and cultivars were made known to the public, they now provide information on what was possible or likely during specific times. They also provide valuable insight into the process by which tastes and vogues evolved. Unfortunately, the value of such material and the need for its preservation has not been understood until recently. Many of our best sales-catalogue collections are stored under poor conditions, inadequately catalogued and improperly preserved. Every effort must be made to assure the survival of this material. It is hoped that the role these sources played in this study will make readers aware of their significance and importance.

In the chapters that follow, after addressing in general some of the major characteristics of the era that affected horticultural matters, a chronological approach is utilized to discuss the three distinct periods: mid-Victorian, late Victorian, and early twentieth century. Chapters III and V consider each annual individually, discussing briefly its history in cultivation, and then, in greater depth, its subsequent development through the Victorian and post-Victorian years. The specific genera considered are: *Abronia* (sand verbena), *Callistephus* (China aster), *Celosia* (cockscomb), *Clarkia*, *Collinsia* (Chinese houses), *Eschscholzia* (California poppy), *Gaillardia*, *Gilia*, *Impatiens* (balsam), *Lathyrus* (sweet pea), *Petunia*, *Phlox* (Drummond's phlox), *Portulaca*, *Reseda* (mignonette), *Tagetes* (marigold), *Tropaeolum* (nasturtium), *Viola* (pansy), and *Zinnia*. Their development is traced through popular magazines, gardening books, journals, sales catalogues and other ephemeral publications. Chapter IV, still within the chronological scheme, focuses on the impact of major public events in generating ideas and trends in the use of flowers. The prominence given annuals in public displays at major exhibitions—such as Philadelphia's 1876 Centennial and Chicago's 1893 World's Columbian Exposition—and the effect of countless fairs, shows, and flower trials popular in the last quarter of the nineteenth century had an accelerating effect upon the refinement and diversity of many annuals.

The text is followed by two appendices. The first, a collection of four complete lists of annuals offered by seedsmen during the 1830s and 1840s, establishes the character and content of catalogues at the beginning of the period under study for comparison with later lists. The second documents the nature of the changes in each annual through specified intervals of time.

I

SEED PRODUCTION AND DISTRIBUTION DURING AN ERA OF CHANGE

DURING THE NINETEENTH CENTURY, numerous economic and social factors affected the development of garden annuals and the entire seed industry as well. A history of the seed industry during this period reflects these rapid changes. By the mid-1800s, a number of major seed houses were well established in eastern North America. Their illustrated lists contained a wide array of both domestic and imported vegetable and flower seeds. The seed catalogues consulted primarily for this study are a selection from those of the major suppliers of the nineteenth century. But before we examine their contents in detail, a brief background of the seed industry as it pertains to the supply of garden annuals in this country is in order.

EARLY DEVELOPMENT OF THE AMERICAN SEED INDUSTRY

The first seed-dealing venture in America was started in Philadelphia by David Landreth in 1784. The Landreth firm was soon followed by companies in other major cities of the Northeast. Founders of such businesses included Grant Thorburn (New York, 1802), Bernard M'Mahon (Philadelphia, 1806), William Booth (Baltimore, 1810), Joseph Breck (Boston, 1818), Thomas Bridgeman (New York, 1824), Charles Hovey (Boston, 1834), Peter Henderson (Jersey City, 1847), James Vick (Rochester, 1849), T. W. Wood (Richmond, 1879), and many others. Philadelphia remained unquestionably the center of commercial seed production throughout the nineteenth and into the twentieth century. By the early 1900s, more than a dozen firms coexisted in the vicinity of Philadelphia. Two that remain thriving operations today are the Robert Buist Company (established in 1828) and the W. Atlee Burpee Company (which began relatively late, in 1875).[1]

Seedsmen were essentially merchants who centralized supplies from individual raisers and sold both to local firms and to the public directly.[2] Initially, America was dependent upon Europe almost exclusively as a source of flower seed. David Landreth began purchasing seed from the Vilmorin, Andrieux Company of Paris as early as 1795.[3] Statements such as "Seeds obtained principally from the best growers of France, Germany and England"[4] often accompanied catalogue descriptions of flowers. Although this dependency was never fully severed throughout the years prior to World War I, production of a home supply of seed also occurred.

During the second half of the nineteenth century, encouraged by a general move toward independence from European trade and a personal sense of pride in the growth of their establishments, seedsmen strove to raise the quality and quantity of seed production in this country. Furthermore, even though the passage time of transatlantic steamship voyages, in operation from 1839, shortened to a swift two weeks, it was believed by some firms that seed "exposed to the injurious influences of a damp atmosphere during an ocean voyage,"[5] lost viability. David Landreth, whose company motto extolled the fresh and pure quality of his "American Pedigree Seeds," stated in his catalogue of 1886 his dislike for imported seed:

> Foreign seeds at best never have . . . the same vitality and vigor of growth as the hard, dry, ripened American seed, and when the soft, immature crops of Europe are subjected to the damp of an ocean passage, their already diminished vitality falls quite 12–15% additional.[6]

Eventually, the desire for more control over actual production in the field stimulated a healthy competition on both sides of the Atlantic which resulted in higher standards of selection and packaging.

Certain economic factors also stimulated this growing industry. Added governmental duties on imported products in the 1870s created an immense interest in seed growing in America. Books such as Francis Brill's *Farm Gardening and Seed Sowing* maintained that there was "money in the garden . . . [and] also in the seeds which supply it."[7] Although the systematic growing of flower seed profited from the experience of noted European growers, successful results in this country had to develop gradually through trial and error. According to an article in James Vick's *Floral Guide* of 1875, it was only in recent years that "flower seeds were grown in America for the market, and these were of the commonest kinds."[8] Vick went on to state:

> All kinds of seeds cannot be grown with profit in any one country. Some sorts are raised best and cheapest in the moist climate of England or Scotland, others are more easily perfected in the south of France; while, on account of some peculiarity of soil or climate, or special skill and experience, others are only to be obtained in perfection from Germany.[9]

Through the investment of time, money, and skill, James Vick and others with equal determination were able to raise many varieties that were cheaper and often better than those obtained from Europe. By 1875, Vick was growing several acres of verbenas, petunias, pansies, cockscombs, and zinnias, and "a score of other things" in smaller quantities.[10]

Individual firms expanded their acreage into outlying areas away from the growing cities, and imported and domestic seeds were tested and rogued in these trial grounds. In the process, methods of plant evaluation were devised to compare known cultivars (described then as "varieties") with each year's incoming deluge of new introductions. This increasing sophistication within the seed industry greatly affected the development of annuals and other ornamentals.

THE INDUSTRIAL REVOLUTION: GROWTH AND CHANGE IN THE
VICTORIAN AGE

In the broader context, it was a combination of social, cultural, and technological factors as well as improved breeding techniques that both directly and indirectly influenced the development of horticulture. It must be stressed that the changes and developments affecting horticulture were part of a greater force influencing every phase of human life during this period: the impact of industrialization. The revolution in technology migrated to America from England during the early 1800s and produced rapid and fundamental changes which literally transformed nineteenth-century society. Although all areas of horticulture were altered in some way by this surge of growth, the commercial aspect responded most directly and positively to the increasingly specialized systems of production and distribution.

A nationwide transportation system was constructed between 1820 and 1915, stimulating commercial relationships across the continent. Although roads and canals provided important links, it was the railroad that played the major role in encouraging economic growth by providing cheap and rapid transportation.[11]

Streamlined packaging was another innovative, labor-saving feature which, in company with low postal rates and the lack of prohibitive interstate restrictions, made it possible for seedsmen to distribute their mail-order products cheaply to all regions of the country.[12]

The growing network of transportation systems, which made possible a broad distribution of goods, naturally helped to destroy cultural barriers. At the same time, a revolution of equally profound nature was penetrating the thoughts and desires of the new mass market. All forms of horticultural literature, from the purely scientific to the strictly popular, were now reaching the public by new methods of communications.[13] Advances in the medium of print greatly affected the nineteenth-century world. The economical manufacture of paper from wood pulp, along with the invention of the steam-powered printing press in 1811[14] and, later, the rotary press in 1875, enhanced the dissemination of printed material.[15] Newspapers, tracts, broadsides, magazines, and books became the major vehicles of communication. Nationwide advertising brought together producer and consumer with a speed previously impossible.[16]

In many ways, ties with England became even stronger during the nineteenth century. There evolved an atmosphere in which ideas and influences flowed in both directions across the Atlantic.[17] Essentially, Britain and America shared a period of history measured by the sixty-four years of Victoria's reign, from 1837 to 1901. It was a period during which a new urban society absorbed not only America's rural and small-town societies, little changed since colonial times, but also Britain's premodern, Old World culture into a society dominated by cosmopolitan ideas. Victorianism describes a world beyond the rule of a foreign monarch. The age was characterized by middle-class standards and values, the products of an industrialized and modernized society.[18]

The Victorian frame of mind combined an enthusiasm for technology and progress with an intense preoccupation with order and proper values. What distinguished Victorians above all else was their "seriousness," a condition believed to indicate their sense of moral urgency and "need for psychological stability amidst the rapid changes occuring during the nineteenth century."[19] As a result, Victorian expression was notoriously didactic. Through the printed media, Victorian writers sought to shape the quality of life. "Art for art's sake," as we know it today, was far from being the principal mode, especially during the mid- (or "high") Victorian period from the 1850s through the 1870s.

Biblical rhetoric and moral instruction permeated all levels of popular literature; articles literally preached the virtues of growing flowers. This familiar tone was dispersed throughout the literature on gardening, as evidenced in the following extract from Thomas Meehan's *Gardener's Monthly* of 1872:

> Flowers stimulate industry as well as lighten toil. For we must have them. We are cold without them, but to have them requires patient study, patient culture, and untiring determination.[20]

Not only would one's personal character be uplifted by the occupation of growing flowers but also one's reputation and social image! Subliminal lessons of integrity were taught in the effort behind tidy arrangements of simple flower beds. And, it was always considered the best practice to raise only a few plants "of choice character and perfect growth [rather] than to have ever so many which are imperfectly developed."[21] Most advertising for annuals contained such messages in one form or another.

As a result of higher living standards and wide-spread literacy, an enormous new audience of middle-class readers was the target for this prescriptive writing. A growing "cult of domesticity" conceived the home as an orderly and secure place where the family could concentrate upon the socialization of the children.[22] Books such as Joseph Breck's *The Young Florist* were written as tools to instruct children and to establish within them "a love of nature and taste for the beautiful that would go with them through life."[23]

Along with woman's place in the home, the Victorian image of womanhood included her role in the planning of her flower garden. In Meehan's "Hints for January" in 1872 he remarked:

> It is a very nice winter study for ladies, and one which in England engages the attention of everyone, from Queen Victoria down, to arrange in winter the beds, and the flowers to fill them, for the summer decorations of the garden. . . . This practice has been gradually growing in England for the past 30 years, until now it is the universal winter employment of all ladies of taste; and to this great interest in flower-gardening by the English ladies, is the present high state of the flower-gardening department there to be mainly traced.[24]

Whether it depicted the lady of the house handing her plans to a gardener for execution or the country wife instructing her husband to take time to prepare the soil for flower beds, the literature of the day created a perception of women and their place in the garden to which the seedsmen catered. This image was to go through a number of transitions as the century progressed. Women writers themselves, from Jane Loudon to Gertrude Jekyll, became more influential, even though they were still dominated by their male colleagues. As further chapters will reveal, women played a crucial role in making the connection between the concepts of garden design and the flowers that were available to fulfill their design requirements.

Finally, the Victorian age fostered a great enthusiasm for the general pursuit of science. New attitudes evolved, and, with Charles Darwin's *Origin of Species* (1858) and other new ideas about the process of change in the biological world, people gradually began to accept the theory that life was in a continuous but understandable state of flux.[25] The selection of superior or unusual cultivated varieties had been going on "half unconsciously for centuries" and had resulted in the permanence of many "types or races of cultivated plants."[26] However, an awareness that this phenomenon was not totally governed by chance began with the discovery in the late eighteenth century of the sexual nature of flowers and the process of pollination.[27] This knowledge, along with Darwin's concept that "the key is man's power of accumulative selection; nature gives successive variations; man adds them up in certain directions useful to him,"[28] served to extend the powers of cultivators still further.

Ironically, though the basic workings of inheritance were revealed in 1865 when Gregor Mendel delivered his paper "Experiments in Plant Hybridization" before the Natural History Society of Brünn in Czechoslovakia, Mendel's discoveries passed unnoticed until 1900, sixteen years after his death.[29] Hybridization, therefore, was not generally carried beyond the first generation and the potential for enormous variety in future progeny was discarded. Even so, plant-breeding methods were decidedly more scientific in the nineteenth century than previously. In this country, Luther Burbank would emerge as a renowned plant-breeding "wizard" by the turn of the twentieth century.

Both for the pleasure of the work and for the profit involved, amateur and professional breeders contributed a deluge of novelties yearly to meet the demands of an expanding market and changing fashions.[30] Horticultural novelties reaffirmed the need for progress so essential to the Victorian. The increase and diversity of cultivated varieties broadened the limits of horticulture. Liberty Hyde Bailey was one of the most invigorating figures on the American scene during this era. In *The Survival of the Unlike*, a collection of essays and addresses first published in 1896, he recognized the implications behind the popular acceptance and desire for rapid advancement by stating:

> This uplift in the common understanding of the science of cultivation, and of the methods of crossing and skillful selection, is extending a powerful accelerating influence upon the variation of cultivated plants. But the most important and abiding evolution is that of the man himself.[31]

Attention to all these permeating influences is essential for an understanding of how the times affected the development of annuals and how annuals, in turn, reflected these characteristic features. This was true not only for their specific forms, but also in the manner in which they were popularized.

II

THE ROLE OF CULTIVATION, COMMERCE, AND TASTE IN THE DEVELOPMENT OF ANNUALS

ANNUALS, AS A BROAD CLASS OF PLANTS, can be defined on several levels. In order to determine what they meant to the average nineteenth- and early twentieth-century American interested in cultivating flowers, certain general concepts which evolved during the period must be clarified. This chapter will consider annuals in relation to evolving horticultural classifications, commercial techniques, and garden styles. These have had an effect on the meaning and use of annuals that has continued to the present.

DEFINING ANNUALS

Annuals . . . blow and die the year they are sown.
(William Cobbett, *The American Gardener*, 1821, paragraph 334)

Among the choicest flowers scattered over the earth are some whose lives are short. They sprout from seed, grow, flower, produce seed in their turn—and then die, all within the limits of a single season of our gardening year.
(Alfred C. Hottes, *A Little Book of Annuals*, New York, 1925)

True annuals, in the botanical sense, are plants that when germinated in spring will bloom the same summer and ripen their seed by the end of that growing season. For practical and ornamental reasons, the horticultural definition must also take into consideration the quality of a plant's flowering performance. Because many plants attain full growth and achieve their showiest blossom display during the first season, they are horticulturally termed annuals even though, in their native habitat, they may actually be biennials or perennials.

Greenhouses, hotbeds, and improved forcing techniques extend the length of the growing season and thus broaden the scope of annuals still further. Plants such as pansies, considered here as annuals, can therefore be listed as either annuals, biennials, or perennials. Cultural methods are, in turn, subject to climatic factors in different regions, which likewise alter the length of the growing season. Therefore, our definition of cultivated annuals must make allowances for the various conditions under which they may be grown.

In the early 1800s, British methods for categorizing annuals were used as models in attempts to create working lists of annuals for American gardeners. Initially, the British arranged annuals in essentially alphabetical lists, the method of such early cataloguers as John Gerard (1596) and William Lucas (1677).[1] By the second half of the eighteenth century, this traditional system had given way to the more sophisticated one of identifying and grouping annuals by their cultural needs. This method subdivided annuals into hardy, half-hardy, tender, and greenhouse or indoor-forcing types. Sometimes a separate section was also given over to climbing annuals.[2] This more precise system of listing annuals was based on the opinions, observations, and documented cultural practices of successful British propagators.

The following descriptions of the major categories appeared in the Flanagan and Nutting *Catalogue* of 1837:

> *Hardy Annuals* which may be sown in open Borders from the middle of February to the end of April.
>
> *Half-Hardy Annuals* which should be sown in March, under hand glasses or on a very moderate Hot-Bed, and transplanted into the Border in the middle of April or beginning of May.
>
> *Tender Annuals* which require more than one Hot-Bed to bring them to perfection, should be sown during the months of February and March.[3]

These brief instructions were undoubtedly quite sufficient in a country where gardening expertise was well established. Adapting such information to the American environment was an issue often confronted in the literature throughout the nineteenth century. Although the mild and favorable English climate was cyclical in its own way, the British environment in general had few of the same difficulties experienced in this country. This situation was lamented by American garden writers such as Peter Henderson, who blamed seedsmen for not helping to solve such problems. His assessment of the situation is evident in the following passage from *Gardening for Pleasure*:

> Our seed catalogues are nearly all defective in not giving more specific directions for the culture of annual plants. If the space for the description of form and color were devoted to telling the time and manner of sowing, it would be of far more benefit to the amateur buyer; but nearly all follow the English practice of giving descriptions of varieties only. There the necessity for such information is less, the people being better informed as to flower culture, and the climate is also more congenial for the germination of most seeds.[4]

These remarks held some truth for the many lists issued by seed merchants who were interested simply in transferring imported seed, with imported descriptions, directly to their customers. But it must be stressed that at that time it was beyond the capabilities of most individual growers and distributors in this country to assemble such information, given the regional variations and lack of centralized data on hardiness and soils. Nevertheless, many influential seedsmen attempted to offer accurate information on the culture of annuals through garden calendars and floral guides as well as through their catalogues, even though regional biases were unavoidably implicit.

A leader of this movement was Bernard M'Mahon, an Irishman, who, after immigrating with his wife to Philadelphia, worked with David Landreth & Company until he was able to establish a successful plant business of his own. M'Mahon is well known for his correspondence and association with such prominent individuals as Thomas Jefferson and for the role he played in respect to the collections of the Lewis and Clark Expedition. William Darlington made the following statement in reference to M'Mahon's work in a letter dated 15 June 1857:

> To him we are mainly indebted, among other favors, for the successful culture and dissemination of the interesting novelties collected by Lewis and Clarke, in their journey to the Pacific.[5]

This letter was printed as a part of a "Brief Memoir" in the eleventh edition of M'Mahon's book *The American Gardener's Calendar*, published twenty-nine years following the author's death in 1828.[6]

From its first publication in 1806, M'Mahon's guide was used as a standard garden reference by many American households. Part of its success was its credibility. M'Mahon sought always to write from his own gardening experiences in the Philadelphia area instead of "referring to works of foreign countries differing materially in modes of culture from those rendered necessary here by the peculiarities of our climates, soils and situation."[7] His ambitious undertaking naturally fell short in its application to the entire country. As a standard reference his work was very useful and widely read, but in upper New England and the deep South, with their vastly differing seasonal conditions, it could be used only as a general guide.

William N. White's *Gardening for the South*, published in 1858, addressed specific growing conditions in the South although most of its information was adapted from G. W. Johnson's *Kitchen Gardening*, an English work. Like M'Mahon, White maintained that all of his knowledge was based on experience and observation, but that his reliance on certain English works was acceptable because of great parallels in climate.

> Our seasons differ from those of the Northern States, in heat and dryness, as much as the latter do from those of England . . . [and] our climate is much like that of the south of England. Hence, while the calendars of operations, in works prepared for the Northern States, seldom agree with our practice, those in English works are often found to coincide with it . . . but . . . the long, dry summers, and still milder winters, of this climate, often

render necessary a peculiar mode of performing the same.[8]

These "long, dry summers" were a major factor in the geographic distribution of certain flower "crazes" such as the sweet-pea phenomenon that lasted from the 1890s through the first decades of the twentieth century (considered in depth in Chapter V). The southern seed merchants T. W. Wood & Sons of Richmond, Virginia suggested fall planting for this crop and offered their collection of sweet pea cultivars with the following warning:

> In the South they have not been received with the same favor as further north, on account of the difficulty to make them bloom well, the hot weather being injurious to their growth.[9]

The calendars and guides of the nineteenth century relied heavily on common sense and trial-and-error methods. The publication of accurate references was an even greater challenge in the more extreme regions. A passage from D. W. Beadle's *Canadian Fruit, Flower, and Kitchen Gardener* represents the approach taken by sources of this vintage.

> We have endeavored to make a selection of those [annuals] that will best repay care and culture in our Canadian climate. It is useless to grow everything. Not even everything that is pretty is worth the requisite labor, when compared with results just as easily obtained by judicious selection.[10]

ANNUALS ACQUIRE NEW SIGNIFICANCE IN THE TRADE

It must be remembered that the growing of flowers, up until around the middle of the 1800s, was considered quite a luxury. When it came to the expenditure of labor, "the average American had little time or taste for gardening until well into the nineteenth century, for his country was new and other more practical needs demanded attention."[11] This situation permeated the way people thought and affected the way the relatively frivolous pursuit of growing flowers was justified and encouraged in the literature. For example, M'Mahon's directions to sow tender annuals "in February in hot-beds with the cucumbers or melons"[12] implies that flowers took second priority to the growing of food crops. A similar attitude was maintained by William Cobbett, an Englishman, who, after living on Long Island for a time, published an interesting collection of observations in *The American Gardener.* The following excerpt on sweet pea and balsam illustrates the point:

> Pea (Sweet)—sown and cultivated like the common garden pea. They should have some sticks to keep them up.
>
> Balsam—sow when you sow Melons, 4′ apart. It will blow early in July, and will keep growing and blowing till the frost comes, and then, like a cucumber, it is instantly cut down. I have seen Balsam in Pennsylvania 3′ High, with side branches 2′ Long.[13]

Cobbett's comparisons not only evoke vivid images, but also suggest the methods which writers of his time used to communicate with their audiences. Everyone knew what a cucumber looked like after the first hard frost, but a balsam was something less familiar.

Nasturtiums (*Tropaeolum majus* and *T. minus*) received quite a bit of attention in garden calendars by reason of their use for food. It is significant, however, that although descriptions of the nasturtium's culinary uses were foremost, the beauty of their flowers was still mentioned. M'Mahon considered the climbing nasturtium "very deserving of culture as well on account of the beauty of its large and numerous orange colored flowers, as their excellence in salads and their use in garnishing dishes."[14] White also mentions the ornamental features of the nasturtium even though his calendar primarily concentrates on agricultural subjects. *Tropaeolum majus*, he acknowledges, has flowers which are "a rich, brilliant orange and continue all summer," but then he adds, "and if not so common, would be thought very beautiful."[15] This final remark almost negates the nasturtium as an ornamental because it was so often found in gardens. However, by mid-century the nasturtium experienced a major shift, moving in the seed catalogues from the vegetable to the flower section, and reference to its culinary uses almost disappeared.

Flower-seed departments acquired new significance as seedsmen recognized the increased demand and started to compete among themselves. David Landreth, whose early catalogues focused primarily on agricultural seed with only a bare listing of flowers, offered considerable sections of annual flower seeds during the latter half of the century. Likewise, James Vick, who was a leader in the movement to create "a taste for the beautiful in gardening, and a true love of flowers, among the people" offered "Annuals and other Plants that Flower the first Season . . . [as] the first and most important section of our Catalogue of Flowers."[16] By the last quarter of the nineteenth century, virtually all seed companies followed this trend.

Henderson's 1881 *Handbook* divides the annuals into two basic categories and describes them as follows:

> *Hardy Annuals* are those which require no artificial aid to enable them to develop, but grow and flower freely in the open air.
>
> *Tender Annuals* are generally of tropical origin, and should not be sown in the vicinity of New York until the first week in May. Indeed, the best rule for all sections of the country, from Maine to Florida, is not to sow the tender kinds until such time as the farmers begin to plant Corn, Melons and Cucumbers.[17]

Henderson's planting guide suggests the increase in leisure time experienced by a large portion of the population at the end of the century. Whereas M'Mahon made no distinction between those who, of necessity, cultivated flowers among the cucumbers or melons when time allowed and those who grew them for their own sake, Henderson indicated quite clearly an urban and suburban clientele distinct from the rural farmer. The very development of the food trade itself was instrumental in the spread of urbanization and the decline in growing one's own food out of necessity.

Ultimately, the seed trade defined annuals in terms of the middle-class homeowner. In 1898, a Boston seedsman claimed that "Annuals are pre-eminently the flowers of the people. They are easily raised, quick in blooming and inexpensive."[18] This advertising technique, which boasted a product "for the millions," was an essential ingredient in marketing strategies for a wide range of commercial entities beyond the seed industry. It is still applied in many contemporary seed catalogues.

POPULARIZING THE NOMENCLATURE

It is important for modern interpreters to understand the absence of standardized formats for plant names and notations in all types of horticultural literature during this period.[19] Because there was no system for proper registration of plant names during the nineteenth century, those used by breeders and seedsmen were arbitrarily ascribed and often inaccurately transferred. Catalogues were notoriously inconsistent in the use of botanical names and cultivar epithets in Latin form. In many cases, the name of a flower was its ornamental description in Latin form, for example, as *Phlox Drummondii rosea alba-oculata* for a pink Drummond's phlox with a white eye or center. Within the same list, on the other hand, might be a cockscomb designated simply as *Celosia cristata* Yellow Dwarf. Such juxtapositions seem confusing when taken out of context. However, if the lists of the 1860s through the 1870s and into the 1880s are compared, the mixtures of Latin and common, or "fancy," names reveal a gradual transition from Latin to an almost exclusive use of common names.

The cultivar names, including Victoria aster, Quaker City mignonette, Glasgow Prize cockscomb, and Crystal Palace Gem nasturtium, reflected significant events, places, or individuals of the time. Today, cultivars can often be found which resemble the descriptions of older forms, but most of the period names—names ascribed during specific periods—are lost.

From a commercial standpoint, the shift from the use of Latin to common names created new avenues for marketing. As seedsmen tried to reach a broader audience, the arbitrarily used and carelessly spelled Latin names became cumbersome, perplexing items in their catalogues. Eventually, the many novelties entering the market yearly, with their attention-catching, "fancy" names, competed with the established cultivated varieties. It is also quite likely that the cultivars of one firm could be made to appear different from those of another through the use of differing fancy names.

In addition, the chances for new cultivars to occur with greater frequency improved as sophisticated breeders consistently grew large quantities of a single type under highly fertile conditions.

As Bailey observed in *The Survival of the Unlike*, this situation especially "encouraged" mutations in annuals, for the yearly propagation of plants from seed tends to cause plants to constantly "change or differ from their parents, and finally to pass so far away from them that they receive new names."[20]

Therefore, with the increased number of novelties resulting from more concentrated and systematic breeding programs, use of common names became a necessity not only for more compelling advertising but also because the Latinate names were no longer practical. Seedsmen and breeders drew from a richer source of descriptive possibilities as they popularized the names of their latest selections. Naturally, problems resulted as seedsmen renamed cultivars either unintentionally or for their own commercial gain. Such practices ultimately created the need for the International Code of Nomenclature.

THE ROLE OF ANNUALS IN SUCCESSIVE GARDEN STYLES

In nineteenth-century America, as elsewhere, "flower gardens went through many permutations brought about by changing fads and fancies."[21] There were several successive styles in the second half of the century that directly affected the popularity of annuals and regulated to some degree their use in the garden. These major trends were: (1) carpet bedding, (2) mixed borders, (3) a composite style of both, and, at the beginning of the twentieth century, (4) a naturalistic style emphasizing harmony of color. The decorative manner in which annuals were used, therefore, creates an added, and perhaps more significant, dimension to their meaning.

In Colonial gardens, where function dominated aesthetics, there were few opportunities to imitate the well-established gardening styles of continental Europe and Britain. However, as leisure time increased by the mid-nineteenth century, Americans were more receptive to changing styles from abroad. It was during that time that the knot gardens, parterres, and shrubberies of the eighteenth and early nineteenth centuries, which depended heavily upon hardy herbaceous plants to provide structure and form to their design, were replaced by what John Claudius Loudon referred to as the "changeable flower garden."[22] This new fashion relied on large reserves of plants that could be "plunged in the borders as wanted."[23]

THE SUMMER FLOWER-GARDEN.

Heddiwiggi. These may be varied to suit the fancy of the possessor with the newest annuals described in the Catalogue, selecting them according to colors, and height of growth. All the beds should be edged with box or thrift. The extent of ground is thirty-two feet square.

For more artistic and complete grounds, we add two plans from two of the most elegant flower-gardens of England.

The first plan (No. 3) is extensive and elaborate in design, and evinces artistic skill and arrangement of a high order. The length of the garden is a hundred and sixty feet, and the width seventy-two feet. The walks are of gravel, and the beds are all edged with box. It may be filled with bedding-plants or with annuals; and, supposing the amateur to desire a mixture of the two, the following is an appropriate list, Scarlet Geraniums and Verbenas being the most effective of bedding-plants:

1. Verbena (blue).
2. Verbena (white).
3. Pansies, of the fine showy sorts.
4. Portulaca (white).
5. Tom Thumb Geranium.
6. Verbena (striped).
7. Portulaca (golden).
8. Campanula Carpatica, with Tree Rose in the centre.
9. The same.
10. Tom Thumb Geranium.
11. Portulaca (white).
12. Verbena (striped).
13. Portulaca (golden).
14. Pansies of the fine showy sorts.
15. Verbena (white).
16. Verbena (blue).
17. Ageratum.
18. Heliotrope.
19. Tom Thumb Geranium.
20. Verbena, Sunset (rose).
21. Portulaca (golden).
22. Portulaca (scarlet).
23. Same as No. 8.
24. Geranium, Lucia Rosea (pink).
25. Tom Thumb Geranium.
26. Tom Thumb Geranium.
27. Geranium, Lucia Rosea (pink).
28. Portulaca (scarlet).
29. Tom Thumb Geranium.
30. Heliotrope.
31. Verbena, Sunset.
32. Portulaca (golden).
33. Ageratum.
34. Same as No. 8.
35. Vase, or Statue. If a vase, to be filled with Verbenas, Petunias, &c. If a statue, to be surrounded with a circle of Oxalis Floribunda.

But, when it is intended to be filled with annuals, this may easily be done by substituting Candytuft, Alyssum, Eschscholtzia, Lobelia, Agrostemma, Petunias, Dwarf Convolvulus, Clarkias, &c.

The last plan which we give (No. 4) is a copy of the flower-garden of the Duchess of Bedford, at Camden Hill, near London. In harmony of arrangement, it stands very high; and, offering as it does a great variety in the disposition of the beds, it contains, in an eminent degree, the two great elements of a select garden, — harmony and variety. "Two things," says a well-known writer, "are necessary to the beauty of a flower-garden, — harmony and variety. Harmony consists in agreeableness of form, likeness of size, and relation of color: variety is the indefinite diversity of vegetative existence. If there is variety merely, the garden is strange, extraordinary, fantastic; it is not fine. If harmony alone is displayed, then it is monotonous, dull, and wearisome. But in the happy combination of the two resides its power to awaken agreeable sensations, and impart delight. This union is well exemplified in this plan."

No scale is given; but we suppose the ground to contain a circle of one hundred feet, — about fifty feet to the inch. The plants employed, annuals and bedding-plants, would be as follows, according to the numbers: —

"The Summer Flower Garden," *Washburn and Co's . . . Guide.*

The designs that these beds represented were the result of "bedding out" plants to resemble carpets, mosaics, or ribbons, creating what has commonly become known as carpet bedding. The essence of this grand style is the use of bold color and uniform habit of growth to achieve a floral display. In 1859, Robert Thompson of the Royal Horticultural Society noted that worthy annuals for this type of bedding were not available prior to the 1830s. However, at that time a change occurred, as many of the hardy, herbaceous perennials began to be replaced by "the bloom of annuals."[24]

Carpet bedding as a phenomenon occurred at the same time as the development of brightly colored cultivars and the technological advances in glass houses that made it possible to grow tender plants indoors in pots. But, did the fancy for such plants develop because they were suddenly available or did they become popular because, as Mariana Van Rensselaer argues in *Art Out-of-Doors*, "public taste had begun to demand bright-colored and stiff material for a special gardening purpose"?[25] Certain trends in the types and forms of plants used appear to bear out her argument.

Initially, in England, many of the annuals selected for carpet beds were those recently introduced into cultivation from North America by David Douglas and other plant explorers. England's Jane Loudon writes, in the *Ladies Magazine of Gardening*, of sowing in the spring such California annuals as *Gilia* tricolor, *Collinsia* sp., and *Clarkia* sp. Even in England, however, these short-lived flowers required several sowings and were still not durable enough to withstand the summer heat.[26] To maintain the broad sweeps of color that this new style demanded, sturdier plants were necessary. In 1861, Thomas Meehan defined in his *Monthly Magazine* the qualities desired for successful bedding plants as opposed to plants grown for their individual merits.

> But, for bedding purposes, a new and striking shade of color, a free blooming character, neat habit of growth, and power to endure a hot, dry sun, are of more importance; and the energies of our improvers should be devoted to this end.[27]

Annuals exhibiting these features, such as dwarf chrysanthemum-flowered China asters, dwarf balsams, and Tom Thumb nastur-

Terrace garden with carpet bedding. Old photograph (author's).

tiums, were very popular at this time. Eventually, even certain perennial greenhouse plants, such as verbenas, coleuses, begonias, and Tom Thumb geraniums, became synonymous with the bedding-out system. It seems, then, that Mariana Van Rensselaer was right in assuming that the style preceded the plant type and created the market for bedding plants.

Many American garden writers embraced the bedding-out style wholeheartedly and promoted it above all others. Peter Henderson is one example of an extreme devotee. As a seedsman with "acres of greenhouses" he was hardly disinterested in promoting this style for commercial gain. In his books he consistently maintained that the mixed border of herbaceous perennials, annuals, and shrubs was a "promiscuous" style which could not equal the grand effect "obtained by planting in masses or ribbon lines."[28] His ideals were taken from the grounds of London's Crystal Palace and Paris's Jardin des Plantes, observed on his travels. Moreover, Henderson was a major critic of American public and private flower gardens. In his estimation, the "monotonous" shrubs of New York City's Central Park, so near his own production facilities in New Jersey, could not compare with the floricultural feats in Britain and Ireland.[29]

Photographic and pictorial evidence from the 1870s and 1880s indicates that there were fine examples of carpet bedding in America, but primarily confined to a few large, private estates and to great exhibitions such as the 1876 Centennial in Philadelphia's Fairmount Park. The more modest attempts of the average citizen were generally seen by American and foreign critics as completely unnoteworthy. In a letter published in *Meehan's Monthly* of 1861, a French correspondent openly condemned the types of plants Americans preferred in their gardens, including "such a worthless flower as petunia . . . [which] looks weedy, has no shading in its color, no luring perfume, nothing at all to recommend it."[30] He concluded that there was a basic difference in attitudes between the French, who gave their gardens first priority, and Americans, who were more interested in the quality of their interior carpets than of those on the grounds.

A Natural Border. William Robinson, *Gravetye Manor*, London, 1911.

"A Wild Flower Garden," *W. Henry Maule's Seed Catalogue*, 1906.

R. Morris Copeland's *Country Life*, published in 1859, also points out the inferior quality of American gardens of his time, but the reason cited is more tangible.

> A real flower-garden is rarely seen in America, and I do not know of any one which can be at all compared either in style, keeping, or size, with the remarkable gardens of England, where twenty acres have been devoted to the display of annual, perennial, and bedding-out flowers. The cost . . . is enormous.[31]

Ultimately, since fiscal realities determine what methods of gardening are feasible, even given the marked increase in leisure time in nineteenth-century America, the cost of labor and the hours required to execute elaborate carpet bedding were prohibitive. The practice of creating these vivid masses of flowers and foliage, though seductive to the public, was actually "the costliest and most troublesome which can be adopted for the adornment of a garden, either large or small."[32]

By the mid-1870s, a subtle trend away from this gardening vogue, which highlighted a few plants to the sacrifice of many, can be detected. Although the bedding system was to retain its supremacy for many more years, the discussion of cultivating flowers for their individual beauty, as distinct from "mere massing for effect," filtered into garden literature. In Thomas Meehan's opinion, published in an 1872 issue of his magazine, the reemergence of hardy plants and the mixed border was a case of history repeating itself and was to be encouraged not only because it gave more pleasure, but also because "it costs less."[33]

William Robinson, the renowned British garden writer, was a leader in the movement away from the bedding system. His many books and his monthly magazine, *The Garden*, relentlessly opposed the "unnatural" beds of England and Europe. Thomas Meehan became an American advocate of Robinson's views, as his book reviews indicate:

> Mr. Robinson is the apostle of a new move in the British Garden, and we trust the influence of his good taste will reach America. We have been too long the imitators of the new fangled rush for long arms of mere color; ribbon gardening . . . has usurped nature; the Englishman

was persuaded that unless his eye was blinded by long vistas of scarlet he had no garden at all.[34]

Today, Robinson's beliefs appear less extreme than they did at the time. Influenced by the native flora of England, which he closely observed during his many trips through the countryside, he proposed that these natural associations be applied to the flower border. Although he used many bold tropical plants, such as bananas, he intended that they be grown in such a manner as to appear as if no human intervention had occurred.

Robinson might be seen as a threat to the use of any annuals in the garden whatsoever, for he was firmly intent upon reestablishing hardy herbaceous plants in the mixed border. However, his attitude was generally favorable toward annuals when their use was in keeping with his concepts. Robinson's ideal garden, as opposed to the "wild garden," was the English cottage garden, which permitted a certain artifice and, of course, the use of annuals. In the chapter on annuals in the 1901 edition of his book *The English Flower Garden*, he recognizes their necessity in the following passage:

> Whatever we may do with perennials, shrubs, or hardy bulbs, the plants in this class [i.e. annuals] must ever be of great value to the flower-gardener; and among the most pleasant memories of flower-garden things are often those of annual or biennial plants: tall and splendid stocks in a farmhouse garden on a chalky soil, seen on a bright day in early spring . . . Snapdragons on old garden walls, and bright Marigolds everywhere . . . Sweet Pea hedges, and Mignonette carpets. . . . However rich a garden may be in hardy flowers or bedding plants, it is wise in our climate to depend a good deal upon annuals.[35]

This nostalgic attitude, with reminiscent perceptions of an earlier, simpler age prior to the industrial revolution, was manifested in the Arts and Crafts Movement which originated in England with William Morris and the pre-Raphaelites and which migrated to America by the end of the nineteenth century. The Colonial-revival period of the late nineteenth and early twentieth centuries occurred in conjunction with this movement and perpetuated the same attitudes.

"Ridgeland Farm," a Naturalistic Garden. Louise Shelton, *Beautiful Gardens in America*, New York, 1915.

Indeed, by the turn of the century the old-fashioned, or grandmother's, garden came into vogue. Robinson's philosophy was echoed by Thomas Meehan, Charles Sprague Sargent, and Liberty Hyde Bailey among others. They encouraged the cultivation of flowers "for their own sake," either in informal beds or as cut flowers. Reference was often made to sweet pea and zinnia hedges. Annual phloxes, mignonettes, clarkias, China asters, and zinnias were grown in beds described simply as the "reserve" or "slip garden."[36]

A love of wild flowers ensued which both echoed this nostalgic urge to preserve the country's original aspect and served as a premonition of the negative ecological effects of the industrial revolution. The following excerpt from an article by Bailey in 1902 suggests this trend.

> To most persons the wild flowers are less known than many exotics which have smaller merit, and the extension of cultivation is constantly tending to annihilate them. Here, then, in the informal flower-border, is an opportunity to rescue them. Then one may sow in freely easy growing annuals, as marigolds, China asters, petunias, and phloxes, and the like. . . . Such a border half full of weeds is handsomer than the average well-kept geranium-bed, because the weeds enjoy growing and the geraniums do not.[37]

Although the naturalistic landscape-revival style eventually took supremacy over the formal bedding system, both schools persisted in various forms and have continued to the present. Furthermore, as landscape gardening developed into the professional discipline of landscape architecture, the execution of both styles became more sophisticated and architectonic. In Mariana Van Rensselaer's view, the geometrical and the naturalistic styles merged to form the mixed or "composite" style in which the merits and limitations of both were recognized and features of either used where they were most appropriate in keeping with the existing features of a place.[38] Louise Beebe Wilder interpreted this trend in her own fashion. Although the following passage from *Colour in my Garden* does not specifically describe a composite style, it does imply the eclectic nature of early twentieth century gardens.

> It requires some fortitude in this day to express approval of the bedding-out system. It has departed, or should have, with the days of antimacassars and hand painted tambourines, and no one wants this period of terrible and useless ornament to return; yet it seems to me that there are times and places where we may still "bed out" with propriety and even grace.[39]

Throughout this period, Gertrude Jekyll, today considered one of gardening's foremost geniuses, was publishing her works on various aspects of gardening style. In her conversational prose she described artistic techniques in the use of highly selected, but wide-ranging collections of plants. She used annuals both discreetly and brazenly, in light touches or great quantities as she saw fit. She was always aware of the flow of a garden and the artistic principles inherent in the eye's need for surprise and rest. She focused simultaneously on the individual flower and its effect in combination with other plants. Above all, she believed it was a gardener's duty to create beautiful pictures.

> While delighting our eyes, they should be always training those eyes to a more exalted criticism; to a state of mind and artistic conscience that will not tolerate bad or careless combination or any sort of misuse of plants. . . . It is just in the way it is done that lies the whole difference between commonplace gardening and gardening that may rightly claim to rank as a fine art.[40]

Jekyll's influence was especially pronounced upon other women garden writers of her day, including Louise Wilder who made constant reference to "Miss Jekyll's" ideas. Louisa Yeomans King, in *The Well-Considered Garden*, believed Jekyll's *Colour in the Flower Garden* was second in importance only to Bailey's *Encyclopedia* for any gardener's library.[41] Color harmony and arrangement became the predominant issue throughout the garden literature of the early twentieth century. Annuals were carefully selected for the visual effect they provided in the mixed border. Gardens became a personal expression of taste combined with a sensitive interaction with nature.

We can see, then, that annuals weathered the change in gardening ideals in the late nineteenth century, passing from being the favorites of the carpet bedders to respected members of the cot-

A Composite Garden. Charles Henderson, *Henderson's Picturesque Gardens and Ornamental Gardening Illustrated*, New York, 1901.

tage gardens of Robinson and the floral designs of Jekyll and Wilder. Paradoxically, many of the improvements achieved by the carpet-bedding enthusiasts may, in fact, have recommended annuals to opponents of that style. In any case, the significance of annuals in the garden went through a number of transformations following upon the tastes and styles of the times.

In summary, during the second half of the nineteenth century, annuals became very important components on the American gardening scene. An entire industry evolved around their development, refinement, and production. Marketing techniques brought many new introductions, with memorable names, to the rapidly expanding suburban population. Furthermore, due to the versatile nature of annual flowers, they became integral to both major styles which evolved during this particularly dynamic period in American horticulture.

"Floral Park," *John Lewis Child's Seed Catalogue*, 1893.

III

HISTORY AND DEVELOPMENT OF A SELECTION OF ANNUALS THROUGH THE MID-VICTORIAN ERA

PRIOR TO THE CIVIL WAR, most American seed catalogues were merely lists offering a limited selection of flowers. The number of cultivars available was generally scant, although this would vary depending upon the emphasis of the particular company. Landreth's of Philadelphia, for example, was at that time primarily interested in vegetable seed and agricultural products; it was not until the 1870s that its flower-seed department expanded.

In Appendix I are examples of four lists from the 1830s and 1840s, those of David Landreth (1832), George C. Thorburn of New York City (1838), Flanagan & Nutting of London (1835), and Joseph Breck of Boston (1845). These are included mainly for contrast with catalogues issued later, but it is also interesting to note the similarities and differences among this earlier group. The Flanagan & Nutting list, in which annuals were divided into hardy, half-hardy, and tender categories, is, by far, the most thoroughly classified. In general, the American lists were less sophisticated, as evidenced by the Thorburn list in which annuals were relegated to one section and the Breck and Landreth lists in which the entire selection of flowers was simply alphabetized by genus.

It is also interesting to note how quickly some of the new introductions were available commercially in England and America. Both Flanagan & Nutting and Thorburn offered *Clarkia* and *Gilia* spp. which had just been introduced into cultivation within the decade.[1] Other plants on the Flanagan & Nutting list, from the

David Douglas introductions of the 1820s, included *Collinsia* and *Eschscholzia* spp.[2] Breck's list indicates that these species were standard items by the mid-1840s.

By mid-century, catalogues offered greater numbers of cultivars, although the numbers may in many cases be misleading. Cultivars listed under different names in the various seed catalogues could easily have been duplicates of one another, for their differences are often not made clear. In Appendix II, some of the cultivars with multiple names are indicated.

In this chapter, selected annuals are discussed individually, beginning with those longest in cultivation and continuing through those introduced during the early nineteenth century. Each section includes a brief historical sketch of the annual in cultivation as well as its status within the seed industry from around 1865 through the mid-1870s. During this ten-year period, Victorian influence was at its height. With regard to flowers, this taste was evident in an intense interest in double-flowered annuals with compact habit that could be used as bedding plants. Dwarf China asters, balsams, and cockscombs were used in this style. The more refined and formal forms of chrysanthemums, dahlias, roses, and camellias were very fashionable, and the double flowers of such annuals as China asters, balsams, zinnias, portulacas, and clarkias were often compared to them. Other general traits developed in annuals during this period included enlargement of the flower and refinement of growth habit.

OLD-TIME FAVORITES AND EARLY NINETEENTH-CENTURY
INTRODUCTIONS FROM ABROAD

This section covers eleven genera of cultivated annuals that are not native to North America. By the mid-1800s these were well known to American commerce and were available in a number of cultivated varieties. Their past use associated them with a variety of garden styles, from carpet beds, parterres, and knot gardens to informal mixed borders. The genera longest in cultivation begin the sequence.

TROPÆOLUM MAJUS.

Nasturtium

There are two species of the common garden nasturtium; both are natives of South America. The first to be introduced into Europe was *Tropaeolum minus*, which was brought by Spanish settlers from Peru in the fifteenth century.[3] This species bears small yellow flowers splashed with a dark orange spot; each of the five petals terminates in a prominent point and the upper one forms a nectar spur. Some of the first plants of this species grown in England were from seed sent to John Gerard by Jean Robin, gardener to the king of France. Gerard, in his *Herball*, writes:

> The seeds of this rare and faire plant came first from the Indies into Spaine, and thence into France and Flanders, from whence I received seede that bore with me both flowers and seede, especially those I received from my loving friend John Robin of Paris.[4]

It was known as the yellow larkspur or Indian Cress, and by the beginning of the seventeenth century Parkinson noted that "It is now very familiar in most Gardens of any curiosity."[5]

Not until the late 1600s was the more vigorous *T. majus* introduced.[6] Because of its climbing habit, it quickly became popular in cottage gardens, trained against old apple trees or across rustic arches and stone walls. The most popular use of this plant was the pickling of the seeds and flower buds in vinegar to be eaten as a condiment with meat. The leaves were also used in salads, like watercress.

A third species, which reached Britian from Peru around 1810, is the canary-flowered nasturtium, *T. canariensis* (referred to now as *T. peregrinum*).[7] This annual climber, with deeply-lobed leaves and bright yellow flowers having fringed upper petals, was strictly an ornamental and was not commonly sold in America until the mid-nineteenth century.

Both *T. majus* and *T. minus* were introduced to America by Bernard M'Mahon in 1806 and were quickly transported across the continent by the early settlers. The common orange-scarlet form was probably most familiar to Americans at that time. The two species cross readily and breeding experiments in England led to the development of the Tom Thumb cultivars by the 1850s.[8] More brilliantly colored forms, including a dark ruby red, were discovered in the wild by such British plant hunters as Thomas Lobb. These new colors were bred into the bushy Tom Thumb cultivars and were on the market by 1857.[9]

The King of Tom Thumb nasturtium was quite well known by the 1860s, and Charles Hovey writes in *The Magazine of Horticulture* for 1866 that he imported seeds of this scarlet flower with brilliant dark foliage from the seed farm of Messrs. Carter and Company, near London.[10] Crystal Palace Gem, with sulphur yellow flowers spotted with maroon, was another of particular note in this series.

Although the Tom Thumbs did not supersede the climbing nasturtiums, they were quite popular. Probably the most significant reason for their success was their dwarf, bushy habit which suited them for use in carpet bedding. Added to this series of compact plants with bright flowers in the yellow to red range was a form with variegated foliage introduced in the early 1870s. This nasturtium was also recommended for use in ribbon and pattern beds.[11]

CELOSIA AUREA PYRAMIDALIS.
CELOSIA. Nat. Ord., *Amarantaceœ*

Cockscomb

A pan-tropical flower long in cultivation, the cockscomb was introduced into Europe around 1570, probably from Asia.[12] The earliest form introduced was the dark red *Celosia cristata*, described in *Hortus Third* as a plant originating from cultivation, with four chromosomes, resembling the wild type, *C. argentea*, with enlarged crested, plumed, or feathered spikes. Paxton's *Dictionary* of 1868 lists *C. cristata* and its varieties *compacta* (red), *elata* (red), and *flavescens* (yellow) as 1570 introductions.[13]

Today's classifications combine the crested, plumed, and feathered sorts under one species, *C. cristata* with a number of subgroups such as the Childsii, Nana, Spicata, and Plumosa.[14] Through the nineteenth century, however, these cultivated races were treated as distinct species. By the mid-1800s, three species were generally recognized in the trade: *C. cristata*, with crested flower spikes; *C. pyramidalis*, with pyramids of feathery spikes; and *C. spicata*, with layers of plumed spikes. Each species was composed of several cultivars.

The cockscomb is well documented in colonial gardens and was most likely fancied for its peculiar appearance and shock of color in a time when gardens had few exotic ornamentals. In Europe there was a great interest in growing cockscombs in pots for floral exhibitions and table decorations, but in America they were generally treated as plants for the open ground throughout the nineteenth century.[15] Nevertheless, American advertisements for the various sorts of cockscombs were often accompanied by instructions on pot culture. Often, the message was that cockscombs performed better in the gardens of this country than in Britain. In Charles Hovey's opinion,

> Such, undoubtedly, is the reason why the magnificent Celosia, now under notice, has not come to us with testimonials of its beauty. The Iresine and the Coleus were the cynosure of all amateurs abroad before they reached us, yet in real decorative effect they can neither compare with the Celosia.[16]

A similar conclusion was reached by James Vick, who remarked when offering the new Japanese Cockscomb in 1873 that it "seems to like the American climate and soil."[17] Customer reactions were mixed on this matter and one letter to Vick bitterly complained:

> The seed you sent me for Japanese Cockscomb produced common things, with pale green leaves, and a coarse top-knot of a brick-dust color, as like your picture as a Shanghai rooster is like a hummingbird.[18]

Correspondence or testimonials were common items in nineteenth-century seed catalogues. Letters from the customers can still be found in modern catalogues, but it is interesting to note that, in the nineteenth century, complaints were printed along with praise.

Cockscombs have always had a controversial appeal, favored by some, despised by others. During the height of the bedding-out craze, they were generally deemed worthy components in ribbon designs. Furthermore, they were seen as desirable substitutes for some of the more difficult-to-grow plants in carpet beds in Britain and elsewhere.

TAGETES SIGNATA PUMILA (full size of the flower).

TAGETES. Nat. Ord., *Compositæ*.

Elegant free-flowering plants, with pretty foliage; very effective in mixed borders; succeeds best in a light rich soil. Half-hardy annuals.

1392 **Tagetes Signata Pumila.** (See cut.) An elegant new dwarf variety, about one foot high; and, when full grown, the plant will measure two feet in diameter, forming a beautiful compact bush, completely covered with flowers, and continuing in bloom until hard frost sets in. Recommended as one of the most showy plants for borders and dwarf beds yet introduced; of the easiest culture. Plants should stand at least two and a half feet apart 10

Marigolds: African, French, and Signet

Although native to Mexico and Central America, the African and French marigolds acquired their common names from the circuitous routes they took before reaching England. Both *Tagetes erecta* (African) and *T. patula* (French) are believed to have reached Spain by the early sixteenth century. The French marigold, known then as the Rose of the Indies, was introduced into England in 1573 by Huguenot refugees.[19] The African marigold, it is thought, became naturalized along the North African coast prior to its introduction into France, and finally into England in 1596.[20] This species took its name in allusion to Emperor Charles V's campaign to free Tunis from the Moors.[21]

Both species were used by the French in their elaborately designed parterres, or, more accurately, knot gardens. In the early seventeenth century, John Parkinson wrote in reference to the African marigold that "this goodly double flower is the grace and glory of a Garden in the time of his beauty."[22] By the eighteenth century, *T. patula* had attained the distinction of being a florist's flower and a favorite with exhibitors.[23]

The first marigolds cultivated in this country were calendulas or pot marigolds, *Calendula officinalis*. True species of *Tagetes* were available by 1806 at the latest, when M'Mahon offered double African, double French, and quilled African marigolds.[24] By the mid-1800s, both species were offered in a number of distinct cultivars for general use in the garden. The greatest objection to marigolds at that time was the disagreeable fragrance which rendered them "useless in hand bouquets."[25]

The marigold most prized for bedding purposes during the 1860s was a new species introduced in the early 1800s. The signet marigold, *T. signata* (now called *T. tenuifolia*), with its dwarf habit and dense mass of blossoms was well suited for carpet bedding. The Pumila group, initially developed in the early 1860s, included various cultivars of extremely compact habit. It was advertised as "a most beautiful plant . . . from 12 to 18 inches in height . . . as round as a ball. The flowers are single, bright yellow, marked with brown. . . . We have counted a thousand on a plant."[26] This exaggeration is technically legitimate because the marigold is a member of the Compositeae, in which a flower "head" is composed of many small, densely packed flowers, but the average customer was probably unduly impressed by such accounts of its floriferous nature.

Garden Balsam

Native to India, Malaya, and China, the garden balsam, *Impatiens balsamina*, was introduced into Europe in 1596.[27] By the eighteenth and early nineteenth centuries, the balsam was well known in American gardens and is mentioned in the lists of Peter Collinson, Thomas Jefferson, Bernard M'Mahon and others.[28] Mixed colors and double blossoms were available at that time and are also on the Thorburn and the Flanagan & Nutting lists (see Appendix I).

The double forms were divided into three groups according to their markings. The camellia-flowered types came in mixed colors spotted with white; the rose-flowered were perfectly double and in solid colors; the carnation-flowered were striped.[29] Nineteenth-century catalogues are confused as to the origin of these

1. BALSAM.   2. PÆONY ASTER.   3. PORTULACCA.

forms: some seedsmen attribute the camellia-flowered balsam to German breeders, others to French. Still others suggest that there is no difference between the camellia and the rose-flowered sorts—that actually the French called them camellia- and the Germans called them rose-flowered.

In any case, these three basic groups were generally accepted in the catalogues of the 1860s and 1870s. Dwarf forms were also available for each group and were recommended for borders and edging. A popular method of growing balsams during this period was to pinch the side branches in such a way that the flowers were not concealed by the leaves. When the more refined balsams were grown in this manner, as for example the popular novelty Solferino (a carnation-flowered type with white blossoms covered with narrow broken stripes and fine spots of red), the entire plant looked like a fine bouquet.

SWEET MIGNONETTE.

Mignonette

Known as the fragrant weed of Egypt, this plant can be traced to the Roman occupation of North Africa, when, it is thought, seed was sent back to Italy from the Adriatic coast. Pliny called it *resedare*, meaning "to assuage" or "to heal," after its use as a sedative and a cure for many disorders.[30] Its Latin name, *Reseda odorata*, derives from Pliny, but the common name is French, meaning "little darling," and refers to its general use during the eighteenth century to perfume the city streets of Paris and "obliterate the offensive odours."[31] The plant was introduced into England in 1752, and shortly thereafter Philip Miller noted in his then recently published *Gardener's Dictionary* that this fragrant *Reseda* "hath been lately introduced into the English Gardens. . . . The Flowers of this Plant have a strong Scent like fresh Raspberries, which will spread over a Room in which two or three Plants are placed; and for this are greatly esteem'd."[32] Mignonette was grown as a pot plant both in France and England during the early nineteenth century and it was not until the mid-1800s that it became a standard garden flower.[33]

Naturally, the mignonette would never have found a place in the colorful carpet beds. D. W. Beadle accurately described it as an "unpretending flower, with scarce coloring enough to distinguish the blossoms from the leaves."[34] The mignonette was grown solely for its wonderfully sweet fragrance, described by some as like the smell of fresh strawberries or, as quoted earlier, raspberries. Forms with larger flower spikes and tinted red or lilac were on the market by the 1860s, but as a result of stringent selection for a single characteristic, these forms tended to lose the fragrance, and eventually lost favor. Parson's New White and The Prize were two "novelties" of this period.

The literature indicates that mignonette was ubiquitous in Europe and England during the nineteenth century, grown by the wealthy as well as the lower classes. Mignonette was also propagated on a large scale by the florist industry as a popular plant for cutting. But although its reputation was carried over to America, there is some question as to how widely it was actually cultivated in this country. Nevertheless, mignonette seed was always available in the trade, suggesting that there was at least a moderate demand.

China Aster

One annual whose development and use are well documented on both sides of the Atlantic is the China aster, *Callistephus chinensis* (previously *C. hortensis*). Because its appearance suggests an aster, early botanists considered it as such (i.e. *Aster sinensis* and *A. chinensis*) after it was first received at the Jardin des Plantes and at Chelsea in the early 1730s. The original plant was discovered by a Jesuit missionary, Pierre d'Incarville, in a field near Peking. It was a single form, with two to four rows of purple ray florets and numerous yellow disk florets.[35]

By 1750, blue, white, red, and purple forms were used to "adorn courtyards and parlours" from Scotland to the Rhine.[36] Double forms, developed in France, reached England in 1752. During the first half of the nineteenth century, Germany became the center of seed production and breeding, especially with quilled types. In fact, Germans so dominated the field that when *Callistephus* was introduced commercially in America by M'Mahon it was known as German aster.[37]

China asters were considered indispensable in parterres and carpet beds as well as in the mixed border, intermingled with early spring flowers or among woody and herbaceous plants. In 1865 James Vick remarked,

> No class of flowers has been so much improved within the past twenty years as this splendid genus, and none has advanced so rapidly in popular favor. They are now as double as the Chrysanthemum or the Dahlia, and almost as large and showy as the Peony, and constitute the principal adornment of our gardens during the autumn months.[38]

Like the garden balsam, China aster was divided into several groups according to habit and flower form. The following list indicates the major types, although some catalogues recognized even more.

> *Truffaut's Peonia-flowered Perfection*: a large-flowered variety, having long reflexed petals [actually florets of a composite flower], and in various colors. The flower stalks grow about two feet high.

New Rose: grows to about the same height, the flowers are very double, of several colors, and the petals finely imbricated.

Peonia-flowered Globe: a very early flowering variety, the blossoms are large, of various colors, and the plant of a stout branching habit.

Dwarf-Chrysanthemum-flowered: grows about a foot high, the flowers are large, finely formed, of various colors.

Dwarf Pyramidal Bouquet: producing a great profusion of flowers. The colors are various, and the plant only about a foot high.[39]

Some of the most frequently encountered cultivars of this period were Imbrique Pompon (a globe type with imbricated florets), Cocardeau (a combination of quilled and flat florets), Hedge-Hog (quilled), and New Victoria (pyramidal).[40]

Pansy

Before the second decade of the nineteenth century, the flower known as pansy was but the wild heartsease, or *Viola tricolor*, a native to Britain's meadowlands and hedgerows and today commonly called Johnny-jump-up. Little was done to improve these small, richly colored flowers until 1814 when two gardeners, working at different country estates and totally unaware of each other, began selecting and breeding the wild heartsease almost simultaneously.

Roy Genders gives one of the more colorful accounts of this story in his book *Collecting Antique Plants*. William Richardson, who worked in the garden of Lady Bennett at Walton-on-Thames, began selectively breeding the flower following the advice of a Hammersmith nurseryman who recognized some notable wild forms on the estate. The second gardener in this scenario was William Thompson, who worked less than ten miles away for Lord Gambier at Iver in Buckinghamshire. Thompson, too, sought to improve upon a number of forms of the wild species. What is significant is that both gardeners had apparently obtained from outside sources an all-blue viola which they used in their crosses.

The combination of the characteristics in these two flowers resulted in numerous refined forms of *Viola tricolor*.[41]

One of Thompson's first seedlings sported this blue coloring and was illustrated in *The Floricultural Cabinet* for May 1835. It was a bright yellow, broad-"faced" flower distinctively edged with sky blue. He named it Beauty of Iver. Four years later, Thompson again discovered an unusual pansy, this time by chance as he was walking along a patch of long-neglected heather. This stray seedling, the first blotched pansy, was named Thompson's Medora.[42]

The tremendous interest in blotched pansies that ensued led to the founding of the Hammersmith Heartsease Society in 1841 and the Scottish Pansy Society four years later. Amateur breeders across Britain began working with the plant, and criteria for show pansies were soon developed. The societies defined the show pansy as a flower with a circular bloom "with a white or cream ground or band to the lower petals (the center portion being covered by a large blotch), the two upper petals being of the same colour as the ground."[43] The show pansy was later to be divided into two sub-groups: the margined (bicolored) and the self colored.

What is known as the fancy pansy developed in Belgium and quickly displaced other pansies in its appeal to the public. In the fancy pansy there is no restriction as to color. By definition,

> The blotch, of violet or chocolate colour should almost cover the whole of the three lower petals with the exception of a wide margin which may be of any colour or of more than one colour. The top petals need not be the same colour as that of the margin of the lower petals and may be rose, cream, gold, purple or intermediate shades. The eye should be bright yellow and clearly defined.[44]

The growing of pansies in this manner by amateur florists and professional breeders was definitely a British and European phenomenon. In America, however, pansies were exhibited at state fairs and horticultural shows, and seedsmen often took such opportunities to display plants raised from their latest shipments of imported seed.

Foreign pansy seed was widely available on the American market and was often sold with detailed instructions for fall and early

A GROUP OF FANCY PANSIES.

spring sowing. The distinction between show and fancy pansies was rarely articulated, and from the descriptions of the cultivars it is difficult to determine if those offered in America would be considered premium forms by pansy connoisseurs. Some of the self-colored forms, like King of the Blacks, Cliveden Purple, and White, were probably of show-pansy stock. Others were more varied and mottled in color: Marbled Purple, Odier, and Striped & Mottled, for example. Vick attributes their origin to Germany but does not indicate whether they are truly fancy forms.[45]

PEAS, FLOWERING—*Continued.*

SWEET PEA.

Sweet Pea

Franciscus Cupani, an Italian monk and amateur botanist, was the first to introduce the sweet pea from the wild into England in 1699 when he sent seed from Sicily to Dr. Uvedale, a schoolmaster at Enfield Grammar School. The weedy-looking plant produced small flowers borne in pairs on short stems. The original flower had purple to reddish-purple standards (upper petals) and light bluish-purple wings.[46] Its sweet fragrance was its most distinguishing aspect, evoking its Latin name *Lathyrus odoratus*.

Robert Furber, "gardiner at Kensington," and others offered sweet peas for sale by 1730. A few years later three distinct forms were known: a purple, a white, and a popular reddish-pink and white bicolored flower known as Painted Lady that had a particularly sweet fragrance. By 1793 scarlet and black-purple forms were added, and in 1817 a striped cultivar appeared on the lists. It was not until 1850 that the first sign of selection for size rather than color occurred in a "New Large and Dark Purple" cultivar offered in England by the Messrs. Noble, Cooper, and Bolton.[47]

The James Carter firm, outside of London, was first to offer Blue Hybrid, in 1860. This was a white-flowered cultivar with a well-defined blue edge, and it is considered the forerunner of the picotee types. Its fine qualities were achieved through selective breeding. Blue Hybrid was highly regarded and received a Royal Horticultural Society First Class Certificate in August 1883.[48]

The next sweet pea to appear (and the first to actually receive recognition from the Royal Horticultural Society—in 1865) was Scarlet Invincible. It was first distributed in England in 1866 and was available in America by 1870 at the latest. James Vick took a special interest in sweet peas and kept abreast of all the new introductions from England. He was probably one of the first to introduce Blue Hybrid and Scarlet Invincible into the American trade. Peter Henderson's firm in New York was another which offered Scarlet Invincible along with Striped, Painted Lady, Purple, and White in the early 1870s.[49]

At that time, sweet peas were not commonly seen in American gardens. They were sold to be grown for a garden hedge or screen, supported by common pea sticks, and for cutting for bouquets. Vick tried to "encourage the general culture of this sweet flower" by offering large papers (or packages) at low prices and by the pound and ounce at about cost.[50] It is interesting to note that this offer was made in 1873, the year of a financial panic in America. Advertisements which emphasized seed as an economical investment reflected a general concern for the worth of the dollar.

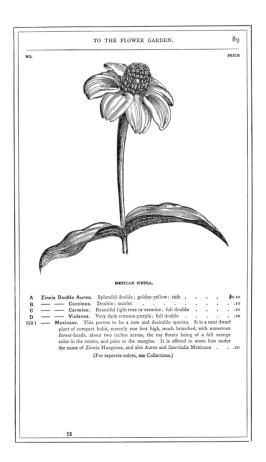

Zinnia

Ornamentally, the zinnia did not have a lot in its favor when introduced in 1796.[51] The wild type, referred to as "medicine hat" by seedsmen, is composed of a single outer row of scarlet ray florets and a central cone of dark brown disk florets. This form persists in nearly all races of modern zinnias, and a few generally appear in any large planting.[52] During the early nineteenth century, the coarse habit and plain scarlet and crimson blossoms of *Zinnia elegans* competed poorly with other, more showy flowers.

It was not until the mid-1800s that the first double forms appeared. M. Vilmorin of Paris received seed in 1859 from M. Grazan, gardener at Bagnères. The plants bore flowers whose central

52

disk florets had developed into showier ligulate, or ray, flowers, similar to those of the dahlia. He offered the seeds in 1861, and Thomas Meehan, upon receiving them, reported in the January edition of his monthly magazine that "more than 50% will come double . . . all colors—from rose to violet amaranth."[53] Meehan's Nurseries exhibited the double zinnia in the fall of that year, but Meehan admitted in his September issue that, of the seeds sent to him "from 'head quarters,' about two thirds came single. This is to be expected from this class of double flowers."[54] Enthusiasm for the double zinnia was still high despite the poor performance of early forms. Meehan himself advised people to "save seed from the doublest and most luxuriant flowers."[55]

Progress was swift, and in 1864, Charles Hovey wrote the following account in *The Magazine of Horticulture*:

> The zinnia is still improving under the hands of skillful cultivators. The first irregularly double blossoms are now brought to as symmetrical a form as the double dahlia, and the original crimson colored variety is sporting into shades of purple, scarlet, orange, and salmon. It is surprising to see how much has been made of this old and neglected plant. But these few changes are only the beginning of what is to follow; for there is no reason to suppose it will not in time give us as great a variety as the dahlia.[56]

Around this time, M. Vilmorin introduced a new species of zinnia which had been discovered in Mexico by M. Ghiesbright. Originally listed as *Z. Ghiesbrightii*, the species was variously known then as *Z. mexicana* or *Z. angustifolia*; its current name is *Z. Haageana*. It was described as differing from *Z. elegans* "in color, as well as in general habit . . . its bushy habit, abundant bloom, and bright golden hue, giving it the highest claims as a garden ornamental."[57] By 1872 a double form was introduced by Messrs. Haage and Schmidt, of Erfurt: *Z. Haageana Flore-Pleno*. It was considered one of the best novelties of that season and quite suitable as a bedding plant for late summer.[58]

Portulaca

This fleshy, trailing annual of the hot and sandy plains of Brazil was introduced into cultivation around 1827.[59] The small flowers were variously colored in the white, orange, and red range, much as they are today. Portulaca's early popularity was not very significant until, as with the zinnia, double forms began to appear. By the mid-1860s, double portulacas were commonly available at up to five times the cost of the single sorts. The "extravagant" price for a carpet of these miniature rose-like flowers was defended by seedsmen because seed of the double forms was so sparingly produced. However, buying the double types was always risky, as Charles Hovey pointed out in 1864.

> The double varieties, like all other double flowers, cannot be relied upon with certainty to produce all double flowers, but the larger part of them will be double, and the single sorts may be pulled up and thrown away or transplanted, unless it is desired to retain them in the same bed with the double kinds.[60]

Portulacas were primarily used for carpeting the ground during the summer months. Indeed, portulaca was viewed as a brilliant, freely blooming annual with limitations. In 1875, Joseph Breck observed:

> No flower exceeds it in the brilliancy of its coloring when opened by the morning sun, and it continues in bloom most of the season; it is good for nothing, only on its bed, being worthless for a bouquet, or other ornamental purposes when cut.[61]

At that time portulacas were sold in separate colors of scarlet, crimson, white, buff, variegated, and yellow. Presumably these were all of the species *Portulaca grandiflora*. Moreover, Vick listed a number of cultivars of *P. alba* and *P. thellusonii* (which was listed in *Paxton's Botanical Dictionary* of 1868 as a variety of *P. grandiflora*). Vick's list of distinct species closely parallels those recognized by Paxton in his *Dictionary* as originating from South America in the 1820s and 1830s.[62]

GROUP OF PETUNIAS.

Petunia

In 1823 a plant with dull, white blossoms, which at night gave off a strong fragrance, was discovered on the banks of the La Plata in South America. Dried specimens were sent to France, where the botanist Antoine de Jussieu constructed the genus *Petunia* and named the plant *Petunia nyctaginiflora* (presently *P. axillaris*). It was soon joined by a second species, with purple blossoms, from Buenos Aires. For a time this second plant was erroneously called *Salpiglossis integrifolia*, and it was first illustrated in the *Botanical Magazine* of 1831 under that name after flowering in the Glasgow Botanical Garden. Eventually, John

Lindley correctly classified it in the genus *Petunia* and named it *P. violacea*.[63]

The common garden petunia, *P.* × *hybrida*, was soon derived from the hybridization of the two South American species. As early as 1837, new cultigens were depicted in colored plates in the *Botanical Magazine*. The flowers varied not only in color, variegation, and size but also in form. Deeply fringed blossoms appeared, and by the mid-1840s the first semidouble forms were described.[64]

Initially, double petunias created quite a sensation. They were produced through a laborious process of hand collecting the pollen from double blossoms, which have anthers but no pistils, and transferring the pollen to the pistils, or female parts, of single flowers which were emasculated. The double form was quite rare and unpredictable, which enhanced its value. However, double petunias were soon found to be of little use except for pots and for cutting. In 1866, Joseph Breck stated that "the double petunias were once the rage, but now, fine, improved single varieties are considered superior."[65] It was recognized at that time that the smaller-flowering, multiflora types were best for bedding.

By 1875, petunias were divided into three classes: double forms; grandiflora forms, including fringed types (produced in a manner similar to the double petunias); and small-flowered, floriferous forms. Each class had a number of cultivars. A few of the more successful bedding cultivars included: Countess of Ellesmere (dark rose with white throat); Kermesina (white with crimson throat); and Inimitable (white with a red margin)—all single-flowered, floriferous forms.

EARLY NINETEENTH-CENTURY INTRODUCTIONS NATIVE TO THE UNITED STATES

The remaining seven annuals are treated as a group distinct from those previously discussed in this chapter. Their introduction dates alone do not set them apart, for the first cultivation of petunias, portulacas, zinnias, and even pansies took place during the same period. The difference lies in the public attitude toward and acceptance of these particular annuals, both in this country and abroad.

English garden writers generally thought of them as the "California annuals" (referring to abronias, clarkias, collinsias, gilias, and California poppies) and highly recommended their use in the mixed border. Literature in this country further emphasized the origins of these wild flowers, and catalogue descriptions often included seedsmen's personal accounts of travels through the West where they encountered these flowers in the wild.

Finally, although they are flowers of this continent, major environmental differences affected the adaptability of some plants to the eastern United States. As this and further chapters will indicate, these flowers met with varying degrees of success as garden annuals. Some adapted easily while others languished and were eventually eliminated from the trade.

Abronia or Sand Verbena

Two species of abronia, *Abronia umbellata* and *A. arenaria* (now *A. latifolia*), generally led the lists of annuals of the mid-1860s. Both are perennial, trailing plants of the California coast, where they grow vigorously in regions where few plants survive. In 1868, Charles Hovey observed that "the abronias are considered charming plants, not unlike the verbenas, with corymbs or heads of sweet scented flowers."[66] These two species and occasionally a third, *A. fragrans* (white with a vanilla fragrance) comprise the extent of the selection of abronias available during the period.

In their native habitat, the yellow *A. arenaria* and the rosy-lilac *A. umbellata* were seen as more robust versions of verbena, and they were advertised accordingly. The following description typifies those of the early 1870s.

> The Abronias are native of California, and in their natural home make a beautiful flowery carpet. The yellow variety, arenaria, delights in the most barren sand hills, and on the borders of the Pacific Ocean, within a few feet of high water, with no other sign of vegetation around, we have seen the clean white sand hills made most brilliant by this pretty plant, which is known on the coast of California as the Sand-Plant. The seed does not always germinate freely, and the plants in some sections do not seem to grow with their native vigor.[67]

Such passages were included in accounts of the Pacific Coast at that time, which were popular because they recalled the adventure of the great movement westward when the country was barely explored. In this context, the plant descriptions were merely a vehicle for a sketch of an exotic region. In any case, the glowing account served to overshadow the unavoidable fact in its last sentence. Their difficult to germinate seeds made abronias almost impossible to grow in areas outside their native habitat.

Clarkia

Named for Captain Clark of the Lewis and Clark Expedition, clarkias are plants of the Pacific Coast as well as the Rocky Mountains. By the 1860s two species were recognized in gardens: *Clarkia elegans*, which has triangular-shaped, lavender-pink petals with long, slender basal claws, and *C. pulchella* with smaller, bright-pink to lavender petals which are lobed.

Meehan reported a "New Double White Clarkia Elegans" offered by Vilmorin in 1861. Unlike petunias, these flowers were not so double as to render them sterile and were therefore not as difficult to reproduce from seed. Many cultivars of both single and double forms were offered by the 1870s in both species.[68]

The English were again more successful with this flower than were Eastern Americans, according to much of the literature at the time. This was likely because the English climate is more similar to that of the Pacific Coast, making adaptation easier. Accounts of London exhibitions in which clarkias received first class certificates appeared in American magazines of the 1860s. Vick wrote enviously of "immense fields ablaze with bright colors, acres each of pink, red, white, purple, lilac"[69] which he encountered in

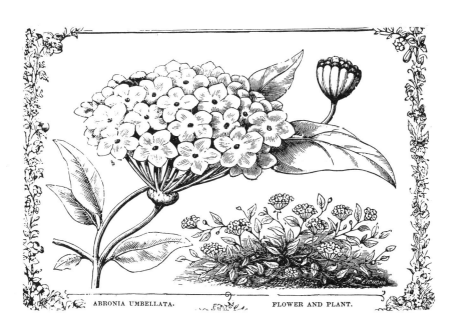

ABRONIA UMBELLATA.   FLOWER AND PLANT.

VICK'S ILLUSTRATED SEED CATALOGUE

CLARKIA.   COLLINSIA.

a country village of Essex. Although, like most seedsmen, he offered a broad selection of cultivars, he readily admitted that "the Clarkia is the most effective annual in the hands of the English florist. It suffers with us in hot dry weather."[70]

Collinsia or Chinese Houses

Zaccheus Collins, a Philadelphia botanist, was honored by the naming of this plant when it was introduced from northern California in 1826. *Collinsia bicolor* (presently *C. heterophylla*), a member of the Scrophulariaceae, has small, irregular flowers with violet or rose-purple lower lobes and white upper lobes.

By the 1860s, seedsmen were offering as separate species several types distinct from the original. These forms would probably be considered cultivars today. Joseph Breck offered several that are not currently recognized or in cultivation, including *C. multicolor* (crimson, lilac, and white) and two varieties of this type, *marmorata* (white and rose marbled) and *bartsiaefolia* (purple lilac). Another species sold by Breck in 1868, *C. grandiflora* (white and lilac) is still recognized, however.[71]

The following excerpt from Vick's 1875 *Floral Guide* again evokes the California landscape:

> The Collinsia is a very pretty, free blooming, hardy annual, that we always liked, but never so well as since we saw it growing wild in California, and which we mistook when at a distance for some new species of Lupin. The marbled, or many-colored, flowers are in whorls of five or six blossoms, and three or more of these whorls on each flower stem. The upper lip of the flowers is white or pale lilac, and the lower one dark purple. About 18″ in height.[72]

GILIA.

Gilia

It is noteworthy that gilias were on seed lists so quickly after their discovery in California and their introduction in 1833. Thorburn offered *Gilia capitata* and *G. alba* by 1838, and *G. tricolor* appeared in early lists as well. The clusters of delicate flowers are borne on freely branching plants of up to three feet high. *G. tricolor*, with fragrant, lilac or violet lobed flowers marked with purple and with a yellowish to white tube, was the most popular for gardens.

The English propagated and exported seed back to the United States and marketed gilias in both countries, primarily as plants for small masses or for cutting. Although highly esteemed in Britain by such garden experts as Jane Loudon, gilias never received a great deal of attention in the American trade.

GAILLARDIA.

Gaillardia

Blanket flower is the common name for this native plant. Its range extends from coastal Virginia to Florida, west to New Mexico and Mexico and north to Colorado, Nebraska, and Missouri.[73] The plant was named for Gaillard de Morentonneau, a French botanist, and the original garden form, *Gaillardia picta* (presently *G. pulchella* var. *picta*) was discovered in Louisiana in 1833.[74]

In America, gaillardias were generally considered good bedding annuals. Their strong, free-blooming flowers met the carpet-bedding criterion for constant bloom throughout the summer's heat and humidity; the plants themselves are vigorous and spreading. The quality of the flower was not as refined as that of such a long-time favorite as China aster; however, Vick noted that, although "the plants are somewhat coarse, and the flowers by no means delicate . . . a good bed of Gaillardia will bring no discredit upon the taste of the cultivator."[75]

By 1875, a double-flowering cultivar probably similar to var. *Lorenziana*, a cultivar offered today (see Appendix II), with all its florets converted to tubular disk florets, was a novelty that generated considerable interest.

Eschscholtzia Californica—2802.

California Poppy

> The early Spanish explorers sailing back and forth along the California coast noted the flame of the poppies upon the hillsides coming down to the sea, and called the coast the Land of Fire. . . . Later, when the Russian expedition of 1815, under Kotzebue, sailed northward exploring the coast, the countless millions of golden cups again won the notice and admiration of the visitors, and Chamisso, the naturalist of the expedition, in reporting the plant, gave it the name of the surgeon of the expedition, Eschscholz, and Eschscholtzia [*sic*] it remains.[76]

Abrams *Illustrated Flora of the Pacific States* confirms this expedition as responsible for the original collections along the coast of San Francisco, and *Eschscholzia californica* was described by Chamisso as a glaucous-leaved, yellow-flowered plant found along the coast.[77]

In its native habitat, the California poppy exhibits a wide variation in characteristics of habit and flower. Some are caused by differences in environment while others are inherent.[78] This situation obviously accounted for great diversity under cultivation. By the 1860s, there were many cultivars available and sold by the seedsmen as distinct species.

Joseph Breck's catalogue of 1868 listed four species of *Eschscholzia*: *E. compacta* (yellow and orange), *E. crocea* (rich orange), *E. alba* (creamy white), and *E. tenuifolia* (primrose with orange center and dark yellow petals); in 1870, *E. aurantiaca*, a deep orange, German introduction was added. All of these forms are currently listed in *Hortus Third* as cultivars of *E. californica* except for *E. tenuifolia*, which is now *E. caespitosa*, a distinct species of central to southern California.

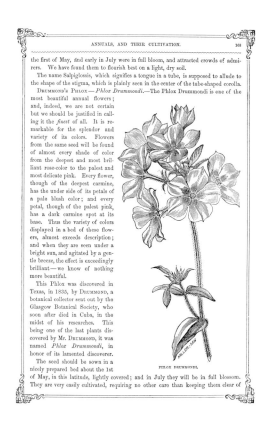

Drummond's Phlox

Naturalist Thomas Drummond discovered this plant in Texas on the second of his explorations of North America during the late 1820s and early 1830s when he traveled throughout the South. Though plagued by violent boils and fevers, he continued to send letters and botanical and zoological specimens to Sir William J. Hooker in Glasgow. Drummond never returned there, but in the last shipment of his belongings, sent in February 1835, were seeds of this annual phlox, which Hooker planted in the Botanical Garden in Glasgow. His description of the plant appeared in Curtis's *Botanical Magazine* of October 1835:

> 3441 *Phlox Drummondii* Mr. Drummond's Phlox. Class & Order Pentandria Nonogynia (Nat. Order Polemoniaceae)
>
> Seeds sent over in the early part of the year 1835 soon vegetated; the plants blossomed most copiously and with equal profusion and brilliancy of colour, whether in the greenhouse or in the open border; and it bids fair to be a great ornament to the gardens of our country. Hence, and as it is an undescribed species, I am desirous that it should bear the name and serve as a frequent memento of its unfortunate discoverer.[79]

This Texas annual became equally popular in eastern North America as far north as Canada. In Beadle's *Canadian Fruit, Flower, and Kitchen Gardener* of 1872 a large section was devoted to the attributes and uses of Drummond's phlox:

> They vie with the Verbena in variety and intensity of coloring, and to be fully enjoyed should be grown in masses of distinct colors. . . . But, planted in any way, whether in separate masses of color, or in ribbons of distinct and various colors, or with all colors indiscriminately mingled, it is one of the loveliest flowers of the garden.[80]

By this time *Phlox drummondii* was available in a great variety of colors, already well established since the 1860s. Violet Queen was considered one of the largest annual phloxes grown in 1868. By 1874, the grandiflora types were issued as a new class of unusual size. Rather than having the white eye and the violet edge of Violet Queen, this form had a large, dark violet eye. This class soon developed many additional cultivars with conspicuous eyes of varying colors.

Of the North American annuals, Drummond's phlox was the most useful for the carpet- or ribbon-bedding style. The many colors available, including white, crimson, scarlet, purple, red, rose, pink, lilac, and violet, were clear and brilliant. Furthermore, the plants grew uniformly and were fairly dependable.

Certainly the impact of the Victorian era can be seen in the types and forms of annuals available in the 1860s and 1870s. There was an obvious emphasis on dwarf, bedding types capable of maintaining a constant and consistent show of color for most of the growing season. Seedsmen sought and highlighted any qualities of annuals that enhanced carpet bedding, such as double forms of portulaca and gaillardia. Flowers that inherently possessed these characteristics, such as the new double zinnias, marigolds, and even balsams, were the stars among the annuals. "Dahlia-like" and "bouquet-type" floral habits were highly prized, and, for this reason, China asters were extremely popular, even though their blooming period is short by modern standards.

It is important, however, not to think of all annuals of this particular period as carpet-bedding stereotypes. The availability of flowers not possessing the above-mentioned characteristics—sweet peas, mignonette, and the California annuals—suggests there was to some degree a curiosity about and a market for flowers grown only for their individual beauty, their fragrance, or, perhaps, even for their novelty. The highly specialized interest in the development of show and fancy pansies spread from connoisseur growers in England and Europe. Although the fascination was limited in this country, there was an awareness of the pansy, albeit somewhat fragmented, on the part of American seedsmen, and a limited selection of improved pansies was marketed in the trade.

IV

The Impact of Public Events

Some consideration must be given to exhibitions, which, like catalogues, books, and magazines, exposed the American public to new plants and styles of gardening. Using the flower shows of Great Britain and the Continent as models, American plant societies, garden clubs, commercial firms, individual plant enthusiasts, and even civic organizations organized horticultural events on local, national, and international levels, especially during the last quarter of the nineteenth century.

The scope of this chapter is bound by two major events in the United States: the 1876 Centennial Exposition in Philadelphia and the 1893 World's Columbian Exposition in Chicago. Both cities' impressive horticultural halls were erected side by side with numerous other buildings, with exhibitions of the arts and the products of industry, forming a collective statement of great achievements. These were international exhibitions, and, not surprisingly, new ideas imported from abroad inspired new directions in the United States.

Horticultural displays at such events were elaborate. They represented a type of gardening well beyond the means of the majority of middle-class visitors. However, these events are important as symbolic models for the countless smaller horticultural shows that cropped up across rural and urban America. This chapter begins with an examination of the floral displays and exhibits at the Centennial and Columbian expositions and concludes with an evaluation of the effect that ongoing shows and trials had on the fundamental development of garden annuals.

MAJOR EXHIBITIONS: THE 1876 CENTENNIAL AND THE 1893 WORLD'S COLUMBIAN EXPOSITION

By the last quarter of the nineteenth century, major international expositions were no longer novel, even in America. Expositions had been mounted in London in 1851 and 1862, New York in 1853, Amsterdam in 1865 (where flowers such as tulips, dahlias, and roses were featured), and Paris in 1867. The 1876 Centennial was unprecedented, however, because it combined a national celebration with an international exhibition, and thus it identified "the Independence and History of America with the Industrial Art and Progress of the World."[1]

Within the conservatory of Horticultural Hall in Fairmount Park were exotic orchids, flowering shrubs, palms, ferns, and other tropical plants. Garden annuals here were few, according to the *Catalogue of Tender Plants Grown at Horticultural Hall*, which lists only *Impatiens balsamina* and *Petunia violacea* varieties.[2] In several rooms leading to the main hall, however, were exhibits of flower seeds, flower stands, gardening tools, and other paraphernalia.[3] These rooms undoubtedly served as display areas and market places for the prominent seed houses. As the *Official Catalogue of the United States International Exhibition* indicates, individual seedsmen also exhibited a variety of merchandise—such as ferns, bulbs, bedding plants, and annuals—in competition.[4]

Surrounding Horticultural Hall were "thirty-five acres of ground, which extend westward over the Belmont road as far as the Catholic Fountain, and which are gay with flowers from all parts of the world."[5] Included in this display was a sunken garden, which contained "beds of flowers, arranged with regard to the pattern and colors in imitation of a carpet."[6] It displayed to the world a perfect example of Victorian taste in gardening at that time. Interestingly, these beds contained none of the common bedding annuals considered in this research. Instead, the beds were composed of "circular plots and straight borders and rings of abutilon [*A. hybridum*], coleus [*C. blumei*], and achyranthus [*Iresine lindenii*] . . . like so many masses of piled velvet."[7] The only other plants in these beds, according to *The Centennial Record*, were

Carpet bedding, Parterre at Horticultural Hall, Centennial Exhibition, Philadelphia, 1876 (photograph: Philadelphia Free Library).

dusty miller (*Centaurea cineraria*), golden feverfew, (*Chrysanthemum parthenium*), and scarlet and pink geraniums (*Pelargonium* × *hortorum* cultivars), although other sources include alternanthera (*A. ficoidea*), stevia (*Piqueria trinervia*), verbena (*V. hybrida*), croton (*Codiaeum variegatum*), and other exotic plants tropical in origin.[8]

These carpets of colorful foliage and flowers resisted the extreme heat and dryness of the summer of 1876 and demonstrated the ultimate perfection of the style. The resilient nature of the plants provided the standard for breeders determined to produce plants for this purpose. To this day, the execution of carpet bedding still aims to achieve the quality displayed in Horticultural Hall's sunken garden. It was one of the best examples of the style in this country for the remainder of the Victorian era.

Certain non-horticultural aspects of the Centennial's theme, however, were in sharp contrast to the ideals that carpet bedding and mechanized gardening represented, and they should not be ignored. The 1876 exhibition appealed to a growing nationalistic spirit in this country, which idealized America's colonial past and brought this trend to new heights. Many contemporary historians attribute the beginnings of the colonial-revival movement to the outstanding popularity of the Centennial exhibition's "New England Kitchen of 1776," along with its exhibits of reproduction furniture and artifacts of the eighteenth century. This trend in the decorative arts would eventually encompass gardening styles as the century progressed. But in the 1876 floral displays at Horticultural Hall the trend was not apparent, although the idea of reestablishing the native environment was vaguely alluded to in a collection of "all the representative trees of this country."[9]

Through the Centennial, the public was directly exposed to new standards and possibilities in gardening as in the other arts and sciences. The juxtaposition of nostalgic and progressive elements actually reflected a recognized tension of the times. Eclectic tastes emerged that ultimately influenced the use and development of garden annuals.

Seventeen years later, Chicago's World's Columbian Exposition, like the Centennial, featured an elaborate horticultural exhibition. Displays within its Horticultural Hall emphasized the

The Lagoon and Island, Columbian Exposition, Chicago, 1893 (photograph: Chicago Historical Society).

development and sophistication of the American seed industry. Actual operations of seed houses were demonstrated, including methods of burnishing and packing seeds for retail trade, methods of testing seed vitality, and methods for growing seeds. One of the more unusual exhibits was a display of vegetables and flowers grown in different latitudes.[10]

Many prominent seed firms participated in the trials for bedding plants and flowering annuals, including Peter Henderson, W. Atlee Burpee, James Vick's Sons (James Vick, Sr. had died in 1882), and David Landreth. In addition to their general exhibits of flower seed, they also featured pansies and grass.[11] Foreign participants included Vilmorin-Andrieux & Co., which exhibited calendulas, marigolds, gaillardias, snapdragons, California poppies, and other flowers in beds along the building.[12]

Certain displays, however, were quite different in nature from those at the Centennial. The focal point of Chicago's floral display was the "wooded island," which the following advertisement describes:

> Fruits and Flowers at the Fair: The "wooded island"— more properly named the flowery island, will be one of the most beautiful and attractive spots at the Exhibitions. . . . There will be acres and acres of flowers of brightest and most varied hues and pleasing perfumes. Little groves of trees, clumps of shrubbery and sinuous walks will relieve this floral display.[13]

This man-made paradise surrounded by water, although still highly contrived, was less formal in character than Fairmount Park's carpet beds of 1876. The descriptions suggest the island was intended to represent a naturalistic treatment of flowers on a grand, public scale.

An even more significant departure from the Victorian artifice at the Centennial was the creation of a naturalistic lagoon, which an article in an 1893 edition of *Garden and Forest* magazine described as "an improvised bit of Nature."

> Rare exotics have not been used, but mainly native plants collected in neighboring woods, fields and swamps. The main object was to secure a massive green effect with cheap home material.[14]

Native plants featured in this setting included willows, iris, red-osier dogwood, elder, sumac, wild verbena, wild sunflowers, bidens, coreopsis, goldenrod, and others.[15]

The presence of such displays at a major and heavily attended exposition undoubtedly publicized the concept of naturalistic gardening throughout the United States. As for garden annuals, these new ideas and modes of gardening affected the use of even the most highly refined, carpet-bedding forms. In Chapter V we will follow in greater detail the changing status of selected annuals as they rose or fell in popularity as a result of shifting tastes.

FLOWER SHOWS AND TRIALS

Whereas the Centennial and Columbian expositions demonstrated the extremes of horticultural technique and development, events on a lesser scale also featured flowers for public appraisal. Numerous plant societies and organizations conducted shows and trials of some form almost yearly, and, in doing so, furthered the development of cultivated plants, including annuals. Accounts of exhibits at state and county fairs and at such organizations as the Massachusetts, Pennsylvania, and New York horticultural societies were written in seed catalogues and guides as well as in the societies' journals. Here the emphasis was less on styles of gardening and more on the improvement of the plants themselves. Commercial acceptance was conferred on annuals receiving awards at these competitions, which were modeled on the highest award at that time, the prestigious Royal Horticultural Society's Certificate of Merit. Because the All-America Selections was not established until 1932, the Royal Society's certificate was the standard for eastern North America as well as for Great Britain.

Following are some of the prominent annuals of the 1880s recognized by horticultural societies. Additional cultivars of the annuals detailed in Chapter III that were introduced during the period from 1876 through 1893 are documented in Appendix II.

Empress of India.

Empress of India Nasturtium

W. Atlee Burpee's catalogue of 1884 quoted the following excerpt from *The Gardener's Magazine* (no date given):

> The flowers of this grand novelty are of a brilliant crimson color, and so freely produced that no other annual in cultivation can approach it in effectiveness, and it would be perfectly safe to describe it as the most important annual in recent introduction.[16]

This new cultivar was as highly esteemed as Carter's Crystal Palace Gem of the 1860s and 1870s. In 1887 it received the Royal Horticultural Society First Class Certificate.

Sheppard's Prize and Trimardeau Giant Pansies

In 1886, Joseph Breck entered a number of newly introduced pansy cultivars in the Massachusetts Horticultural Society's trials. His entries included many of the recently developed French hybrids attributed to three pansy specialists: Bugnot, Cassier, and Trimardeau.[17] These pansies had tremendous blossoms in mixed colors, differing dramatically from the English show and fancy pansies. Cassier's Superb, Roemer's Superb Giant Prize, Belgian Blotched, Bugnot's French, Victoria Red, Trimardeau Giant, Sheppard's Prize, and Breck's International Prize were entered. The award "for the best 50 blooms and also the best six plants in pots"[18] went to Sheppard's Prize and Trimardeau Giant.

These new, giant races of pansies were to become the standards for the twentieth century. Bailey, in his 1933 edition of *The Standard Cyclopedia of Horticulture*, credited the three French specialists for surpassing the pansies grown in England and Scotland. He noted:

> About forty years ago . . . Bugnot of St. Brieuc, and Cassier and Trimardeau of Paris, made immense strides in developing the pansy, and their productions were a revelation to the horticultural world. Such sizes and colors were previously thought impossible. Trimardeau developed a new race with immense flowers and very hardy constitution. His strain, crossed with those of Cassier and Bugnot, has given a pansy which is superseding the older English varieties.[19]

New Plumed Celosia

Burpee advertised this new cockscomb as the "Triumph of [the] 1889 Paris Exhibition."[20] It was considered an improved cockscomb, with bronze foliage and fiery blossoms. Glasgow Prize Celosia, imported in the mid-1880s, was another improved sort with enormous crested, scarlet blossoms.

Eckford and Laxton Sweet Peas

From the time when Scarlet Invincible was awarded a First Class Certificate in 1865, the Royal Horticultural Society began to take particular interest in sweet peas. Some of the most outstanding introductions of the 1880s and early 1890s were those of Henry Eckford of Shropshire, England. He began work on sweet peas in 1876, and by the 1890s dozens of cultivars in the American trade could be traced to him.[21] Thomas Laxton of Bedford, England began to work with sweet peas around 1877 as well.[22] The work of these men, along with the efforts of British seed firms such as Messrs. J. Carter and Co., resulted in an extensive variety of sweet peas which began to enter the American trade by the 1880s.

Joseph Breck's firm in Boston was one of the first to feature these new introductions. First-Class-Certificate winners which he advertised included:

> Invincible Carmine: R.H.S. award winner for 1886; large, intense crimson-carmine flowers.
>
> Orange Prince: R.H.S. winner for 1887; with a bright orange pink standard flushed with scarlet, and bright rose wings veined with pink.
>
> Princess of Wales: R.H.S. award winner for 1886; flowers shaded and striped with mauve on a white ground.[23]

Award-winning status given to such selections was undoubtedly a selling point by the 1880s. In their lists of novelties, seedsmen took full advantage of the desire thus stimulated in the public.

V

DEVELOPMENT OF ANNUALS FROM THE 1890s TO 1914

THE CHARACTER OF SEED CATALOGUES and popular gardening magazines changed dramatically as the nineteenth century progressed to its conclusion. The vividly colored lithographs and fine-lined drawings of mid-century gradually gave way to inferior black-and-white photographs. The photographs did, however, present the scale and habit of plants more realistically than stylized drawings, in which it was difficult to distinguish the blossoms of zinnias from dahlias or balsams from roses.

The tone of the catalogues also changed. By the 1880s, seedsmen, obviously eager to produce their own lines of flower seed, offered named cultivars such as Vick's Branching China Aster, Fordhook Fancy Poppies, or Emily Henderson Sweet Peas. By the 1890s this trend began to assume a more assertive quality with the emergence of the series of "Defiance" flower seeds, an appellation alluding to American resistance to foreign-grown seed. The W. Atlee Burpee Company was a leader in this marketing approach, but other firms, such as Thorburn's, Landreth's, and Breck's, also offered Defiance Balsams, Defiance Pansies, Defiance Largest Flowering Petunias, Allen's Defiance Mignonette, and Breck's Defiance Zinnias. These, accompanied by American flags and other patriotic insignias, were mild manifestations of the nationalistic atmosphere that characterized the period.

This nationalistic trend took on also a nostalgic tone as Americans idealized the colonial past. Grandmother's gardens became the vogue, based on the English cottage-garden concepts initiated by William Robinson. Prominent figures in American horticulture, from Thomas Meehan to Liberty Hyde Bailey and Charles Sprague Sargent, reinforced this longing for a less-complicated age and for the simple, old-fashioned gardens where color and fragrance mingled with abandon. They reacted against progress and modern notions that seemed to discard past values and experiences. Perhaps they felt that in the garden such changes could be arrested, if only temporarily.

The ideals of gardening and the use of annuals underwent yet another change from the previous decades, caused largely by the works of women writers who promoted Gertrude Jekyll's ideas. The use of color in the garden was a common preoccupation, and color nuances were crucial topics in gardening books and early twentieth-century editions of periodicals such as *Country Life in America*, *House Beautiful*, *House and Garden* and many others. Seed catalogues, answering the need for specific colors in garden design, took great pains to separate sweet peas, Drummond's phlox, China asters, zinnias, portulacas, and pansies carefully into color categories.

Annuals, reincorporated into the mixed border, were used in ways differing from those of the Victorian era, but not in entirely new or revolutionary ways. What was different was the vastly enlarged selection of plants available by the beginning of the twentieth century. This final discussion of a selection of annuals from the mid-1890s through 1914 considers the relationship between these more-refined plants and popular gardening ideas.

Semple's White Branching Aster.

China Asters

> August is the month of China Asters. I find many people are shy of these capital plants, perhaps because the mixtures, such as are commonly grown, contain rather harsh and discordant colours; also perhaps because a good many of the kinds, having been purposely dwarfed in order to fit them for pot-culture and bedding, are too stiff to look pretty in general gardening. Such kinds will always have their uses, but what is wanted now in the best gardening is more freedom of habit.
> (Gertrude Jekyll, *Colour in the Flower Garden*, 1908)

Thus, Gertrude Jekyll pronounced a new attitude toward the well-known China aster. She enticed the critics of carpet bedding to use these plants in different ways from the carpet bedders by devoting an entire section of her own garden to a selection of pure violet-purple and lavender China asters. For her, it was merely a matter of choosing the right types, which she selected from the "Comet, Ostrich Plume, and Victoria classes—all plants with long-stalked bloom and a rather free habit of growth."[1]

William Robinson similarly preferred China asters in the flat or reflexed classes; however, he chose from these groups the more medium-sized forms, such as the White Mignon (introduced around 1890) and Hedge Hog. He felt that the very large-flowered China asters, such as Peony-flowered and Victoria, had a tendency to become coarse or to show open centers. These fine lines of taste reflected the author's personal views and should be evaluated with this in mind. However, a general dislike for the dwarf, carpet-bedding sorts was clearly evident during this period. Robinson felt that the dwarf types looked "very well for the short time they are in bloom, but their dumpy habit of growth fits them chiefly for pot work."[2] This attitude, if not so bluntly put, was at least insinuated in *The House Beautiful* in 1902 when the dwarf Triumph asters were recommended because "the flowers are borne freely and openly, so that the plant as a whole is full of grace notwithstanding its diminutive size."[3]

This preference for China asters with a freer habit was also apparent in American periodicals, as, for example, in an article by G. W. Kerr in *The Garden Magazine* in which the "flat rayed" forms are described as the only ones worth growing. This division included the Globe asters (catalogued as Truffaut's, Peony-flowered Perfection, Semple, tall Triumph, and other late-flowering, branching types) and the flat or reflexed group (catalogued as Washington, Mignon, Victoria, Queen of the Market, Crown, and Comet types).[4] The majority of improved forms introduced during this time fell into these groups, which also included Semple and Vick's Branching asters developed by American breeders.

The popularity of China asters would, in the next decade, be threatened not by a shift in taste, but by the devastating effects of two diseases: wilt (caused by a soil- and seed-borne fungus of the genus *Fusarium*) and yellows (a virus spread by a species of leafhopper).[5] These problems were evident during the first decade of the twentieth century and were mentioned in catalogues and periodicals. J. M. Thorburn's 1905 seed catalogue suggested that disease (presumably wilt) was caused by the use of large quantities of manure that was too fresh. His remedy was to stir fresh wood ashes or unslaked lime into the soil.[6] Kerr's article in *The Garden Magazine*, cited earlier, also addressed disease problems. Apparently, they were not considered serious enough to recommend abandoning the cultivation of China asters altogether, but Kerr's only solution was "attention to proper cultural methods."[7] The seed industry was yet to be plagued with the consequences of growing single crops repeatedly in the same soil. The challenge of developing disease-resistant strains of China asters remained for the breeders of the 1920s and 1930s.[8]

Nasturtium

> Common as the *Tropaeolums* undoubtedly are, in the sense that they are known to every one and found in the child's garden in a backyard as well as in the largest public and private gardens in the land, they are not to be despised.
>
> (Charles H. Curtis, *Annuals Hardy and Half-Hardy*, n.d., p. 88)

Both the tall and dwarf forms of nasturtiums were popular in catalogues through 1914, although Lobb's and canary-flowered nasturtiums appeared to decline in importance and some firms discontinued them. An exception was the J. M. Thorburn firm, which offered an extensive list of *Tropaeolum lobbianum* cultivars during the early twentieth century. The problem in analyzing these lists is that some of the forms cited as Lobb's nasturtiums were listed in other catalogues as tall or climbing nasturtiums (*T. majus*). Such confusion was not new in the trade, but was a particular problem with the several species of nasturtium.

Improvements in nasturtiums were generally in the areas of color and size of flower. Of the numerous cultivars that entered the market, Ida Bennett cited Sunlight and Moonlight as the finest introduced in years. She described them respectively as clear golden yellow without markings and pale cream showing wine-colored markings.[9] David Landreth offered them as part of a new strain of nasturtiums from California with "beautiful giant flowers [and] . . . a great range of color, including Sunlight, Moonlight, Twilight, Butterfly and all the California Giants."[10] He sold them as a collection called Landreth's Giants.

Color intensity was a recurring theme in the trade descriptions of nasturtium cultivars. Some of the more popular novelties were forms with variegated leaves or mottled flowers. Inevitably, a bizarre combination of these unusual aspects was offered by Thorburn's as a Variegated-leaved Queen of Tom Thumb Chameleon; the flowers were said to be of many colors on the same plant.

With the revival of old-fashioned gardens, climbing nasturtiums were again valued for informal use, such as trailing over fences, trellises, and stone walls. Mention of their culinary uses also returned in some catalogues. Landreth's 1914 catalogue concluded its page-long list of nasturtiums with the following notice:

> We call the attention of table epicures to Nasturtium sandwiches, the effect being most novel, and to the palate most delicious, both green leaves and flowers being used.[11]

Mignonette

> There are few places in a garden where Mignonette would be out of place.
> (Charles H. Curtis, *Annuals Hardy and Half-Hardy*, n.d., p. 81)

Like nasturtium, mignonette was a favorite in mixed borders. Landreth's sold it as a plant for all situations, "in a box or pot on the window in some narrow, pent-up alley, or in the open ground in the summer."[12] Successive sowings were always recommended in order to provide flowers throughout the season and pot plants over the winter.

Many improved strains were available during this period, including florists' forms such as Allen's Defiance and Machet. However, many felt this effort on the part of the breeders was useless. Gertrude Jekyll, who shared this belief, made the following remarks about mignonette in *Annuals and Biennials*:

> The beauty and true charm of Mignonette lie in its sweet scent and modesty of colouring—the sweetest scented of all is the cheapest sort sold by weight. Where there are so many flowers of brilliant colouring for the adornment of our gardens we may well leave Mignonette alone with its own modest colour and incomparable sweetness. For these reasons the kind called Mile's Spiral is one of the best for the colour is little altered, and it has a fine spike and excellent scent.[13]

Harriet Keeler echoed these sentiments in *Our Garden Flowers* where she concluded that "enlarging the spikes has not always improved the odor . . . [and] the old garden form, with its light, sweet, pleasant fragrance, holds its own fairly among the fifty improved varieties offered by the trade."[14]

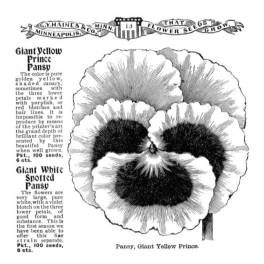

Pansy, Giant Yellow Prince.

Pansy

Enlargement of the flower continued to be the aim of pansy fanciers. The French strains, developed in the 1880s, were quite popular, and improved forms of Trimardeau, Odier, and Bugnot's Superb appeared regularly. Unlike the show and fancy pansies, these strains came in a variety of markings—self-colored, blotched, or mottled. The Odier, Cassier's Giant Odier, and Masterpiece were "five-spotted" sorts with enormous flowers.

Joseph Breck offered some of the better forms and claimed his seeds were grown under contract "by persons who give special attention to their particular pet strain."[15] Mr. Bugnot, himself, was said to have grown the Bugnot's Private Strain and Bugnot's Blotched offered by Breck. Dependence upon European production continued through 1914; however, nearly every seed firm included in this study offered at least one selection of its own strain. These included Burpee's Defiance, Landreth's Philadelphia, Thorburn's Superb, and Breck's Boston Prize, which was advertised as "an American strain . . . superior to any European mixture."[16]

Pansies were commonly treated as flowers for the spring to be discarded and replaced by summer-bedding annuals. For more informal gardens, however, garden writers suggested growing pansies in shady areas all summer.

✣ DOUBLE PETUNIAS. ✣

Petunia

> The tendency now-a-days with most subjects is to get the flowers as large as possible, and the Petunia is no exception to the rule; indeed, it is one in which this character is strongly marked, for the huge lumpy blossoms composed of a mass of flimsy petals are totally wanting in any pleasing feature, while out of doors they are easily spoilt by wind and wet. A pinch of seed will frequently yield plants which are for all ordinary purposes greatly superior to many of the named varieties, whose size is often their only claim to recognition.
>
> (William Robinson, in *The Garden*, August 29, 1891, p. 195)

Seed lists of petunias from the 1890s onward confirm William Robinson's observation, made in an issue of his magazine. Since the mid-1800s, grandiflora petunias in single, fringed, and double forms had been always available. The development of the California Giant petunias during the early 1890s was considered a major triumph for American breeders. This strain was generally sold as a mixture of single, fringed blossoms over four inches in diameter. They were either striped, veined, or blotched in a wide array of colors.

However, the large forms were not always preferred in the literature of this period. Gertrude Jekyll did not even consider petunias in *Colour in the Flower Garden* and she recommended only the white cultivars in *Annuals and Biennials*. Louise Beebe Wilder specifically endorsed petunia Rosy Morn, a small bedding cultivar, used in combination with *Salvia patens*, but she made no mention of other forms in her *Colour in My Garden*.[17] Generally, petunias recommended for reliable performance in the garden were those in the single, small-flowering categories. Often, cultivars specifically cited were some of the oldest, such as Countess of Ellesmere, Blotched and Striped, and Kermesina. Although the giants were the pride of the seedsmen, it appears their use was probably more for pot culture or for cutting than for flower borders.

MARIGOLD, LEGION OF HONOR.

Marigold

> The brilliant orange African Marigold [is] one of the most telling plants of the time of year.
> (Gertrude Jekyll, *Colour Schemes for the Flower Garden*, 1919, p. 88)

Comparison of descriptions and photographs of marigolds at the turn of the century and now shows that their appearance then was virtually indistinguishable from today's. The major forms of African marigold (*Tagetes erecta*) included Eldorado, Nugget of Gold, Lemon Queen, Orange Prince, and two dwarf forms—Pride of the Garden (yellow) and Delight of the Garden (lemon yellow). The favorite French marigold (*T. patula*) from around 1900 through 1914 was Legion of Honor, sometimes called Little Brownie, with golden blossoms marked with maroon or velvety red. Two cultivars of signet marigold (*T. tenuifolia*) were also available, under the names Golden Ring (orange) and Cloth of Gold (yellow).

Marigolds were used to fill spaces between perennials, and, the low-growing French and signet marigolds in particular, for borders and edging. Louise Wilder found marigolds useful for sunny effects, along with zinnias, California poppies, calendulas, mulleins, Mexican prickly poppies, and other yellow and orange flowers.[18]

But it was the pale-sulphur and the brilliant-orange African marigolds in particular that Gertrude Jekyll used to carry the experience of gardening to new dimensions. Her flower borders were designed in sequences of grays and purples that swelled to strong reds and scarlets, blood-reds and clarets, moved to yellows, and then, as the eye became saturated with these rich colors, reverted to grays, blues, and purples again. She appreciated African marigolds for the optical effects they achieved in her color arrangements.

> The brilliant orange African Marigold has leaves of a rather dull green colour. But look steadily at the flowers for thirty seconds in sunshine and then look at the leaves. The leaves appear to be bright blue![19]

Her theory, therefore, was that grays and blues would refresh the eye and appear more brilliant after "the preparation provided by their recently received complementary colour."[20]

Drummond's Phlox

> The annual *Phlox Drummondi* alone has produced distinct varieties enough to furnish a garden with almost every shade of colour.
> (William Robinson, *The English Flower Garden*, 1900, p. 712)

A great deal of enthusiasm was expressed about Drummond's phlox on both sides of the Atlantic during the early twentieth century. American seed catalogues offered a diversity of forms and cultivars far exceeding what is now available in standard catalogues. Much of the seed was produced in what were then the seed-growing districts of the Continent, including the Messrs. Haage and Schmidt's nursery at Erfurt, Germany. An article in the *Gardener's Chronicle* of 1898 described these fields when in full bloom as resembling "an endless Turkish carpet of the most glowing colors."[21] Annual phloxes were divided by the British into seven classes, all of which were represented in American lists, although here they were often combined into three or four classes. The more extensive British classification is described below:

> *Phlox drummondii*: the old tall class, growing around 1 foot in height and distinguished by bright and uniform colours.
>
> *P. drummondii grandiflora*: same height as former . . . produces largest flower trusses of the whole tribe. In this section a good many varieties with a large pure white center (*–stellata*) or with white eye and dark center (*–oculata*). Colours beautiful when cut and looked at but do not show off well in groups.
>
> *P. drummondii Heynholdi*: 10 inches high, grows sparingly. Best for pot culture. Succeeds in hot summer in open ground. Colours mostly vermillion-scarlet to rose.
>
> *P. drummondii cuspidata* and *fimbriata*: star-red and fringed phloxes introduced around 1891. One foot high. Curious looking, lovely margined flowers show beauty only in a cut state, as in a flower bed one may call this class ugly.
>
> *P. drummondii Graf Gero*: eight inches high. Upright growing variety, produces small flower trusses, recommended only for pots.
>
> *P. drummondii Hortensiaeflora*: one of the most beautiful forms. Dwarf, five to six inches high. Forms big round bushes with large flower trusses; the individual flowers almost as large as the *grandiflora* section. There is no better class for the flower garden than this. [Many in this class were listed as *grandiflora* types in American catalogues.]
>
> *P. drummondii nana compacta*: a dwarf, compact class, four to five inches high. Does not produce such large flower trusses as *Hortensiaeflora*, but of a more uniform growth.[22]

Annual phloxes were used in mixed borders, cutting gardens, and in solid masses continuing the carpet- or ribbon-bedding tradition. Some of the dwarf forms were recommended as ground covers, and an article in an 1891 issue of *The Garden* suggested that "beds or groups of standard Roses and other plants may be brilliantly carpeted with them without in any way interfering with their growth."[23] The great variety of colors and their free growth habit worked well in color schemes in the mixed border. For the open garden, Gertrude Jekyll preferred *P. drummondii coccinea* (a *grandiflora* type) and the salmon-colored cultivar called Chamois Rose (of the old, tall class). But she also found the dwarf kinds well suited for rock work.[24] Several annuals discussed here were also popular in the emerging fashion for rock gardens.

DOUBLE ZINNIA.
ZINNIA ELEGANS FL. PL.

Zinnia

It is not long since the Zinnia was considered a coarse plant, with muddy color effects in the flowers, and we must confess that we are surprised at the improvements made in the past few years.
(James Vick, *Garden and Floral Guide*, 1908)

Although the coarse habit of growth was still a question of taste for some, the many new colors available in zinnias (*Z. elegans*) improved their status tremendously. Catalogues offered separate as well as mixed colors in all groups, which consisted of:

> Giant-flowering Double: about 3 feet high double flowers from four to five inches in diameter.
>
> Tall Double: same height as above, smaller flowers.
>
> Dwarf Double: only two feet high; compact growth. large flowers.
>
> Lilliput: fifteen inches high, and thickly branched. flowers but little larger than a daisy in size, very double . . .
>
> Tom Thumb: varies from four to twelve inches in height, and from six to fourteen inches in diameter, forming compact, free-flowering, pigmy bushes, suited for edgings, small beds, and pot plants.[25]

Color selection was an individual matter. In *The Well-Considered Garden*, Louisa King made several references to the "flesh-colored" zinnias which worked best in her designs. For Louise Wilder, the golden-yellow and burnt-orange as well as the ashen-pink and salmon-pink zinnias were considered "good perpetual flowering plants for filling in the blanks left by biennials."[26]

Zinnias could still not be relied upon to come true to color. For the more informal, old-fashioned border this situation was perfectly acceptable. However, for carefully thought-out color schemes, many best-laid plans went awry. Louisa King warned her readers to avoid any zinnia seed marked "Rose," while Harriet Keeler considered zinnias in general to be unpredictable. "No other flower of cultivation takes on such a surprising number of hues, but there is always an element of chance in what a seed may produce."[27]

VAUGHAN'S IMPROVED CALIFORNIA POPPY. (ESCHOLTZIA.)

California Poppy

Interest in this brilliant flower increased during the early twentieth century. Breeders selected for compactness of habit and purity of color. Mandarin, Cross of Malta, Rose Cardinal, and Golden West were the major products of this type of selection available in the trade (see Appendix II for descriptions). A flower with solid crimson petals both inside and out was a goal sought by many, and, in 1906, such a plant was introduced to commerce by Luther Burbank in his Burbank's Crimson California Poppy. A comparable form was introduced around the same time by the Carter firm as Carmine King, and, by his own account, a year earlier as *Eschscholzia californica intus rosea* by Alfred Watkins.[28] In any case, the form available to the American public was Burbank's introduction.

In both America and England, California poppies were recommended for mixed borders, banks (in masses), and edging. Fall sowing was occasionally suggested.

Cockscomb

Catalogues of the early 1900s continued to advertise cockscombs in the predictable, sensational manner of their predecessors of a few decades earlier. The crested forms by that time included Dwarf Chamois (fawn colored), Empress (purple), Glasgow Prize (crimson), Golden Yellow, Queen (rose), Vesuvius (scarlet), and Queen of Dwarfs (dark scarlet). The plumed sorts also came in a number of cultivars, the most noted being Golden Yellow, Thompson's Superb (crimson), and Ostrich Feather (crimson).

Although catalogues endorsed them for bedding, many proponents of more naturalistic gardening discouraged their use. Louise Wilder reacted negatively to many bedding-plant stereotypes and "coxcombs" were included in her list of "taboo plants" for bedding out, along with coleus, alternanthera, castor oil plants, crotons, cupheas, and many others.[29] Gertrude Jekyll was a bit more tolerant. Although she considered the magenta cockscomb to be "a plant unbeautiful both of form and colour,"[30] she did approve of the feathered or plumed kinds as long as the harsh crimsons were avoided.

Balsam

> The old florists had a great love for Balsams, and the flowers were very popular many years ago.
> (Charles H. Curtis, *Annuals Hardy and Half-Hardy*, n.d., p. 51)

Its reputation as an "old-fashioned" flower recommended balsam for nostalgic gardens. Aside from the Defiance and Solferino forms, which were essentially improved strains of the traditional Carnation-, Rose-, and Camellia-flowered forms, new introductions were few during the early twentieth century. The garden literature of the period recommended inclusion of balsams in mixed borders, especially for shady areas, but they were not featured plants in general. Balsams appear to have been reduced to a secondary role as garden annuals.

Gaillardia

Gaillardia

Gaillardia, or blanket flower, became increasingly important in the flower garden, although, as with balsam, there were few improved introductions during this period. Robinson considered the flowers "valuable for their long duration both on the plants and in a cut state,"[31] and Harriet Keeler also considered them excellent as cut flowers because "the heads stand up on good, self-respecting stems and take water freely."[32] Their durability in the garden from midsummer until late autumn rendered both *G. pulchella* var. *picta* and var. *lorenziana* attractive subjects for mixed borders.

CLARKIA.

COLLINSIA.

PORTULACA

Portulaca

The large-flowered single and double forms of portulaca were standard items in any seed catalogue of this period. They were sold to be grown in carpet-like masses on sunny slopes or hot exposed situations with soils prone to dryness, much as they are advertised today. During the early 1900s, mention of their use in rock gardens also began to appear.[33]

Gilia

American catalogues generally offered gilia as mixed colors, however a few offered the three-colored species (*G. tricolor*) in separate white (*–nivalis alba*) and azure blue (*–capitata*) forms. Gertrude Jekyll considered them "pretty plants . . . but not of the first importance."[34] There is evidence that gilias were also used in rock gardens, at least in England, as well as in mixed borders.

Abronia

By the twentieth century abronias appeared less frequently in American seed lists. The species, when offered, remained unchanged, and descriptions were little altered. Both Peter Henderson[35] and William Robinson[36] suggested that these trailing plants were "well adapted for rock-work" or in an open, well-drained border.

Collinsia

Like gilias, collinsias appear to have been of minor importance in the garden during the early 1900s. Peter Henderson found them "of great beauty, and deserving of cultivation, being well adapted for massing and for mixed borders."[37] *Collinsia bicolor* was the type most often sold, although some catalogues did offer *C. bicolor alba* and *C. verna* (sky-blue and white), all of which are now called *C. heterophylla*.

Clarkia

Clarkias were moderately popular during this period, especially with the development of the cultivar Salmon Queen. Louisa King found this salmon flower to be "one of the most graceful and remarkably pretty annuals which have ever come beneath my eye"[38]

905. Gilia tricolor.
Natural size.

and Jekyll, likewise, mentioned that "seed growers have obtained a desirable salmon-coloured variety,"[39] although she still considered the original plants which bore "clouds of pink bloom . . . [to be] the most refined in color."[40]

Tom Thumb cultivars were also available by the 1890s, including a distinct form named Mrs. Langtry which was white with an evenly defined center marking of crimson. The Tom Thumb sorts were developed from *C. pulchella*, whereas the double-flowered sorts were developed from the larger species, *C. elegans*. American seed firms did not offer the diversity of cultivars available in Europe as a rule; however, some of the more outstanding cultivars did appear to be in demand.

Sweet Pea

> The Cult of the Sweet Pea has now extended far outside the Kitchen garden.
> (Charles H. Curtis, *Sweet Peas and Their Cultivation*, 1908, p. 13)

Henry Eckford, a renowned British flower breeder and specialist, began working with sweet peas in the 1870s and was so successful that by the turn of the century over 130 cultivars could be attributed to his selections. Eckford aimed to improve not only the color, but also the substance and quality of the flower. His growing techniques, which included giving each seedling a great amount of growing room, increased the plants' vigor appreciably. The majority of sweet pea cultivars available during the 1880s and 1890s in this country were Eckford introductions.[41]

Joseph Breck and Sons was one of the first American enterprises to recognize and import the improved sweet peas, but W. Atlee Burpee and Company soon became a leader in seed distribution. California became a seed production center during the 1890s primarily through the seed farms of C. C. Morse and Company of San Francisco.[42] It was at the Morse farms that the type of plant from which Burpee's Cupid sweet peas were derived was developed. Cupid was introduced by Burpee in 1893 as the first dwarf, white-flowered sweet pea.[43]

The miniature sweet pea received mixed reactions. Charles Curtis's history of sweet peas, which is naturally biased toward English work, states that the Cupids had not become popular in that country either for pot culture or for the edges of flower borders. His quotation of a description from the 29 June 1895 issue of *The Garden* has a definite Robinsonian quality.

> Sweet Pea Cupid—A miniature variety, of which nine pots were shown, the shoots being about a foot long and profusely flowered, the flowers pure white, and as large as those of the ordinary forms. We fail to see the value of this, which is a poor apology for a noble garden flower, and nothing will be gained by dwarfing Sweet Peas into comparative insignificance. The craze for a pigmy strain of our best garden flowers is to be deprecated.[44]

Nevertheless, the new introduction was granted an Award of Merit by the Floral Committee of the Royal Horticultural Society at the show to which this article referred. In this country, Cupid sweet peas were highly acclaimed.

Other important American introductions of the late-nineteenth and early-twentieth centuries were Navy Blue, America (striped white and scarlet), and Helen Pierce, which was the first marbled form.[45]

The most sensational breakthrough in sweet-pea history occurred at the First Exhibition of the National Sweet Pea Society on 25 July 1901, only a year after the first sweet-pea show—the show in which Eckford entered so many of his introductions—was held at the Crystal Palace. It was at the 1901 exhibition at the Royal Aquarium, Westminster, that Silas Cole, gardener for Earl Spencer at Athorp Park, Northhampton, entered a pink sweet pea, the first known to have frilled or waved segments. Charles Curtis described that day:

> The day was . . . remarkable for the terrific thunderstorm that raged over London; some of the rain found its way through the Aquarium roof, and, bringing with it a portion of London's soot deposit, it stained the papers on the show tables and made a terrible mess. But the day was memorable for another reason; Countess Spencer, a beautiful pink Sweet Pea, with frilled or waved segments was exhibited by Mr. Silas Cole . . . and it marked the beginning of a new era in Sweet Pea development. The Sweet Pea experts unhesitatingly granted the newcomer a First Class Certificate.[46]

The Countess Spencer, which proved to be highly variable from seed, was not "fixed" until 1903 and was then redistributed as John Ingman.

It is interesting that the new form, like the first refined *Viola tricolor*, occurred almost simultaneously in four different localities. Crane and Lawrence's *The Genetics of Garden Plants* traces this occurrence to an unstable cultivar called Prima Donna. Accounts indicate that the waved type appeared in rows of Prima Donna grown by Mr. E. Viner at Frome, Henry Eckford at Wem, and W. J. Unwin at Histon, all about the same time that Silas Cole discovered Countess Spencer. Crane and Lawrence describe this phenomenon as follows:

> The waved character is determined by a recessive gene which must have arisen as a mutation from the dominant normal form. . . . In a very short time the new character was combined with numerous other flower colours by hybridisation.[47]

As the extensive lists of sweet peas reprinted in Appendix II indicate, the response in this country was overwhelming. In an attempt to clear-up some of the confusion that understandably resulted from these endless lists, many seed catalogues separated them into color categories or classes. Selection was still a difficult matter and garden writers advised their readers merely to try a few new ones each year and then decide.

By 1914, the love of sweet peas was at its height. In Landreth's catalogue of flower seeds, the sweet pea section began with an impressive description of this craze.

> The whole world is engaged in a further development of the Sweet Pea—a development as to size, color and stability, or firmness of form. Horticultural Congresses are called together in London, Paris, Berlin, and American cities at appropriate seasons to admire the new forms, and to pass awards of merit to the credit of the successful breeder of the new types.[48]

The note of optimism and the suggestion of unity among the world's leading cities seem particularly ironic in light of the times during which this announcement was made.

Of all annuals popular during this period, the sweet pea was most representative of the taste and spirit of the post-Victorian years. The delicate, airy blossoms and trailing habit could not have competed with the colorful, stocky bedding plants of the previous decades. Sweet peas had an informal quality that appealed to nostalgic sentiments. At the same time, the surge of improved forms satisfied the public's desire for novelty. Like reproduction furniture that flooded the market as part of the Arts and Crafts Movement, sweet peas combined elements of progress with hints of the past. Such developments are never without a price. Gertrude Jekyll subtly alluded to this unavoidable situation in *Annuals and Biennials*.

> It is a curious and extremely regrettable fact that so many of the fine Sweet Peas of the newer kinds are almost scentless. Forty years ago the old hedge of mixed Sweet Peas was the sweetest thing in the garden.[49]

It is significant that so much attention was focused on the delicate-flowered, rankly growing sweet pea during the early years of the twentieth century. Obviously, the times were right for just such a flower to attract interest and satisfy taste. The grand flowers of the Victorian era, the stiff China asters and dwarfed bedding forms of annuals, gave way to those with looser habits of growth. Color schemes in the flower borders captured the interests of the gardening public and women garden writers, in particular, carried this interest to a fine art.

By 1914, many annuals developed in the Victorian heydey were appearing in different forms and were used in ways considerably removed from the artifice of the carpet-bedding vogue. Still, within post-Victorian revivals there were inescapable elements of the previous era. At about this time, Louise Wilder observed the current trends and wrote: "It requires some fortitude in this day to express approval of the bedding-out system . . . yet, it seems to me that there are times and places where we may still 'bed out' with propriety and even grace."[50]

Late Victorian Garden.

VI

SOME GENERAL OBSERVATIONS

THE YEARS 1865 THROUGH 1914 were marked by a unique sequence of social changes in eastern North America that had a profound impact upon the development of annuals. Technological innovations and changing cultural attitudes characterized this pivotal age and acted as driving forces behind the swift advances in horticulture. The change in some annuals was so dramatic that by the first decades of the twentieth century they bore little resemblance to the types familiar to early-nineteenth-century gardeners.

Many of the transformations observed in annuals are attributable to the Industrial Revolution. Improved transportation systems distributed flower seeds swiftly to expanding markets where tastes were quite different from the narrow, elite consumer public of previous years. Annuals were popularized by organized commercial agencies whose marketing techniques broadened the exposure and salability of their products. Popularizing forces were further augmented by the increased availability of gardening ideas and information in an expanding literature that encouraged the use of a variety of flowers by all levels of society. By World War I an astounding selection of cultivated varieties had been produced and marketed by a number of well-established seed firms. The American seed industry achieved substantial independence, although influences from abroad were still apparent.

Gardening styles also dictated which types of plants were most commonly used. Our close ties with England directly affected the cultural attitudes and tastes behind gardening vogues within this country. The ornate carpet bedding of the mid-Victorian era from the 1850s through the 1870s, which featured dwarf, stiff plants with large, brightly colored flowers, exemplifies a gardening style that filtered to America from across the Atlantic. It fostered an interest in annuals of uniform growth that were durable, vigorous, and relatively easy to cultivate. Breeding efforts focused on these qualities and many new "bedding annuals" were developed.

As the Victorian era wore on, a strong eclecticism crept into gardening, reaching its height in the early twentieth century. As a result, garden designs ranging from highly architectural to simple and informal often showed European, English, and even Oriental influences. Anti-Victorian sentiments and national pride at the turn of the century led to stylistic revivals which stimulated interest in traditional "colonial" and "old-fashioned" cottage gardens. But the best gardens, as seen by the arbiters of garden taste, were creative expressions of the designer rather than effusive floral potpourris or imitations of mosaics and ribbons. Annuals were used in combination with herbaceous and woody plants in these more subtle, impressionistic designs, rather than being featured as set pieces in the middle of lawns. Although many of the same plants used in carpet bedding reappeared in these very different gardens, cultivars with freer habits were selected. While bedding out did lose favor, certain sectors of society remained loyal to the style. The tradition of the bedding-out system survives and is still widely used for broad public appeal, especially in parks and exhibitions.

The breeders' increased ability to direct the breeding of plants led to great variation in annuals. Seedsmen announced each season's latest discoveries and breakthroughs with much fanfare, cultivating the public's curiosity and passion for novelty. Even the most bizarre aberrations were touted, and discriminating gardeners were wise to look to the more credible endorsements of horticultural societies, garden experts, and reliable seedsmen to sort out the truly desirable introductions. Again, many contemporary breeding pursuits, as well as the advertising techniques used to promote them, can be traced to late-nineteenth-century origins. The qualities in annuals sought by modern breeders—ease of culture, vigorous habit, purity of color, extended blooming periods—were similarly sought by breeders a century or more ago.

Appendix I

Four Pre-1865 Lists of Annuals

The following reproductions of seed lists are included to establish the extent of annuals available prior to the Civil War and to illustrate the nature of earlier catalogues. They are intended to be used for comparison with lists of the last half of the nineteenth century, details of which are given in Appendix II. The lists included here are:

David & Cuthbert Landreth. *Periodical Catalogue*. Philadelphia, 1832.

Flanagan & Nutting, Seedsmen and Florists. *A Catalogue of Seeds*. London, 1835.

George C. Thorburn. *Catalogue of Kitchen Garden, Herb, Flower, Tree and Grass Seeds*. New York, 1838.

Joseph Breck & Co. *Catalogue of a Choice Collection of Flower Seeds*. Boston, 1845.

PERIODICAL
CATALOGUE
OF
GREEN-HOUSE AND HARDY HERBACEOUS
PLANTS AND SHRUBS,
ORNAMENTAL AND FRUIT TREES,
BULBOUS AND TUBEROUS
FLOWERING ROOTS,
VEGETABLE AND FLOWER SEEDS,
CULTIVATED BY
D. & C. LANDRETH,
AT THEIR
NURSERIES AND GARDENS,
ON FEDERAL STREET NEAR THE UNITED STATES' ARSENAL,
AND FIFTH STREET BELOW FEDERAL.

WITH A LIST OF
Agricultural and Horticultural Implements,
CONSTANTLY FOR SALE AT THE
WAREHOUSE, NO. 85 CHESNUT STREET.

[issued in 1824]

TO WHICH IS ADDED,
A TREATISE
ON THE CULTIVATION OF VEGETABLES IN GENERAL,
Founded upon their own experience.

Philadelphia:
PRINTED BY WILLIAM STAVELY.

1832.

POT AND SWEET HERBS.

> thyme,
> sweet bazil,
> sweet marjorum,
> anise,
> rosemary,
> sage,
> summer savory,
> winter savory,
> carraway,
> lavender,
> coriander,
> pot marigold, &c.

In papers of 6 cts. and upwards, or 5 dollars per hundred, assorted.

EARLY SEED POTATOES, 75 *cents per bushel.*
  white onion sets.
  yellow onion sets
  garlick sets.
also cabbage, lettuce and cauliflower plants, in their season.

FLOWER SEEDS.

ANNUAL BIENNIAL AND PERENNIAL.

t, denotes tender ones. *c*, climbers. Price six cents per paper, or five dollars per hundred assorted.

Flos Adonis, or Pheasant's Eye	*Adonis miniata*
Sweet Alyssum	*Alyssum maritimum*
Love Lies Bleeding	*Amaranthus caudatus*
Straw Colored do.	——— *var. lutea*
Prince's Feather	——— *hypocondriacus*
t Three Coloured Amaranthus	——— *tricolor*
China Aster	*Aster sinensis*
Animated Oats	*Avena sensitiva*
Scarlet Snap Dragon	*Antirrhinum majus*
Double Columbine	*Aquilegia vulgaris*
Rose Campion	*Agrostemma coronaria*
Chinese Hollyhock	*Althea rosea*
Strawberry Spinach	*Blitum capitatum*
t Scarlet Cacalia, or Tassel flower	*Cacalia coccinea*

FLOWER SEEDS.

Starry Marigold	*Calendula stellata*
Venus' Looking Glass	*Campanula speculum*
American Centaurea	*Centaura Americana*
Great Blue Bottle	—— *cyanus major*
Purple Sweet Sultan	—— *moschata*
Yellow ——	—— *suaveolens*
Blessed Thistle	—— *benedicta*
Crimson Dwarf Cockscomb	*Celosia cristata*
Yellow Cockscomb	—— *var. lutea*
Ten week Stockgilly flower	*Cheiranthus annuus*
t Annual Chrysanthemum	*Chrysanthemum coronarium*
Job's Tears	*Coix lachryma Jobi*
Dwarf convolvulus	*Convolvulus minor*
Canterbury Bell, *white and blue*	*Campanula medium*
Scotch Silver Leaved Thistle	*Carduus sp.*
Wall Flower	*Cheiranthus cheiri*
Brompton, Stockgilly flower	—— *incanus*
Elegant Coreopsis	*Coreopsis tinctorea*
Dahlia	*Dahlia superflua*
Bee Larkspur	*Delphinum elatum*
Carnation Pink	*Dianthus caryophyllus*
Pheasant Eyed	—— *plumarius*
Clove	—— *hortensis*
Sweet William	—— *barbatus*
Purple Fox Glove	*Digitalis purpurea*
White —	—— *fl. albo*
Branching Larkspur	*Delphinium consolida*
tc Purple Hyacinth Bean	*Dolichos lablab*
tc White Haycinth Bean	—— *var. albo*
Variegated Euphorbia	*Euphorbia variegata*
c Cotton Plant	*Gossypium herbaceum*
Purple globe Amaranthus	*Gomphrena globosa*
White — —	—— *fl. albo*
Tall Sun Flower	*Helianthus annuus*
Dwarf —	—— *v. nanas*
Bladder Ketmia	*Hibiscus trionum*
White Candytuft	*Iberis sp.*
Purple —	—— *sp.*
Double Balsamine, mixed	*Impatiens balsamina*
t Scarlet Morning Glory	*Ipomœa coccinea*
t Cypress Vine	—— *quamoclit*
t Sweet Peas, various	*Lathyros odoratus*

FLOWER SEEDS.

tWinged Peas	*Lotus tetragonolobus*
Red Lavatera	*Lavatera trimestris*
Rose and blue Lupins	*Lupinus pilosus*
Yellow Lupins	——— *luteus*
cEverlasting Peas	*Lathyrus latifolius*
Honesty, or Satin Flower	*Lunaria annua*
Perennial Lupin	*Lupinus perennis*
Scarlet Lychnis	*Lychnis chalcedonica*
Curled Standing Mallow	*Malva Crispa*
Caterpillars	*Medicago circinnata*
Hedge Hogs	——— *intertexta*
Snails	——— *scutellata*
tIce Plant	*Mesembryanthemum chrystallinum*
tSensitive Plant	*Mimosa pudica*
Marvel of Peru	*Mirabilis jalapa*
Devil in a Bush, or love in a Mist	*Nigella damascena*
Evening Primrose	*Oenother garandiflora*
White officinal Poppy	*Papaver somniferum*
Double Carnation Poppy	——— *fl. pleno*
cScarlet Flowering Bean	*Phaseolus multiflorus*
Red Persicaria	*Polygonum orientale*
tPolyanthus	*Primula polianthus*
tCowslip	——— *veris*
Mignonette	*Reseda odorata*
tWhite Egg Plant	*Solanum melongena*
Purple Rudbeckia	*Rudbeckia purpurea*
Starry Scabious	*Scabiosa stellata*
Wing Leaved Schizanthus	*Sehizanthus pinnatus*
Purple Jacobea	*Senecio sp.*
White ———	——— *sp.*
Lobel's Catch Fly	*Silene armeria*
Sweet Scabious	*Scabiosa atropurpurea*
t cBalsam Apple	*Momordica balsamina*
African Marigold	*Tagetes erecta*
French	——— *patula*
Nasturtium	*Tropæolum majus*
Heart's Ease or Pansey	*Viola tricolor*
Golden Eternal Flower	*Xeranthemum lucidum*
Purple —— ——	——— *annuum*
Mexican Ximenesai	*Ximenesia enceloides*
Red Zinnia	*Zinnia multiflora*
Yellow	——— *pauciflora*
Violet colored Zinnia	——— *elegans*

G

A CATALOGUE

OF

SEEDS,

SOLD BY

FLANAGAN & NUTTING,

Seedsmen and Florists,

9,

MANSION HOUSE STREET,

OPPOSITE THE MANSION HOUSE,

LONDON.

METCALFE, PRINTER, 3, GROCERS' HALL COURT, POULTRY.

1835.

Chicory
Furze
Grass, Sweet Vernal
    Meadow Fox-tail
    Smooth Meadow
    Rough-stalked
    Crested Dog's-tail
    Meadow Fescue
    Sheep's Fescue
    Fine or Heath
    Rough Cock's foot or Orchard
    Timothy
    Rib or Plantain
    Brome or Soft
    Lucerne
    Ray or Bents
    St. Foin
    Yarrow
    mixed, for Lawns
Linseed, or Riga flax
    English
Mangle Würzel
    Yellow
Mustard, White
    Brown

Oats, Black
    Pollard
    Dutch Brew
    Tartarian
    Potato
Peas, Grey Rouncival
    White Boiling
Rape or Cole Seed
Tares, Winter
    Spring
Turnip, Green Round
    White Round
    Red Round
    Globe
    Large Scotch Yellow
    White Tankard
    Red Tankard
    Green Tankard
    Stubble
    Yellow Swedish
    Red-topped Swedish
Wheat, White
    Red
    Spring
    Talavera

FLOWER SEEDS.

HARDY ANNUALS,

Which may be sown, in open Borders, from the middle of February to the end of April.

Adonis, Flos, vernalis
    æstivalis
Agrostemma, Cœli rosa

Alyssum, maritimum, or sweet
Alkekengi

Amethystea cœrulea
Ambrosia, Species
Anthericum annuum
Anagallis Indica
 cœrulea
 new blush
Antirhinum bipunctatum
 medium
 speciosum
 spartium
 latifolium
 triphyllum
 viscosum
 versicolor
Arctotis tristis
 anthemoides
Aster tenella
Athanasia annua
Balm, Moldavian, White
 Blue
Belvidere, or Summer
 Cypress
Bidens diversifolia
Bladder, Ketmia
Briza maxima
Calendula hybrida
 stellata
Calliopsis bicolor
Calochortus venustus
 splendens
Campanula pentagonia
 persicifolia
Candytuft, Normandy
 purple

Candytuft, white rocket
 small white
 sweet scented
 dark purple, new
  variety
Catananche lutea
Caterpillars
Catchfly, Lobel's, red
 white
 New Siberian
Centaurea cyanus, minor
 do. double dwarf
 Crupina, major
 Crocodilia
 Elongata
Cerinthe aspera major
 minor
Chenopodium scoparium
Chrysanthemum carinatum
 (tricolor)
 coronarium, white
  yellow
 do. quilled
 aureum
Clary, red top
 purple top
Claytonia perfoliata
Collinsia grandiflora
 verna
Convolvolus minor (tricolor)
 (bicolor)
 major purpurea
 Michauxii
 fine-striped

Convolvolus, colors, separate
Clarkia pulchella
         alba
     elegans
Coreopsis tinctoria
      diversifolia
      Atkinsonia
Coronilla Securidaca
Corydalis sempervirens
Dracocephalum canescens
Elsholtzia cristata
Echium violaceum
     plantaginium
Eutoca multiflora
Gilia capitata
    pulchella
    tricolor
Glaucium Phœnicium
     Violaceum
Gypsophila elegans
      viscosa
Hawkweed, yellow
      purple
      new straw
Hedgehogs
Horns
Hyoscyamus agrestis
      pictus
Iberis umbellata
   odorata
Impatiens
Isotoma axillaris
Knautea orientalis
Larkspur, Dwarf Rocket

Larkspur, tall Rocket
    fine Rose
    Branching
    Neapolitan
    Unique
    in distinct colours
Lavatera, Red
    White
Love lies Bleeding
Lobelia, Annual
Lupins, Yellow
    White
    large Blue
    do. Rose
    small Blue
    Straw-coloured
    Dutch Blue
Lupinus mutabilis
    micranthus
    Cruikshankii
Lusania calycina
Lychnis læta
    dwarf
    fulgens
Mallow, Scarlet
    curled
    Venetian
Malope trifida
    grandiflora
Malcomia Africana
Madia elegans
Melilotus cœrulea
Mignonette, Sweet
    Upright or Reseda

Nasturtium, tall
    dwarf
    new dark
Nigella, Roman
    Double Dwarf
    Orientalis
    Spanish
Nolana paradoxa
    prostata
Œnothera grandiflora
    Lindleyana
    purpurea
    rosea
    sinuata
    tetraptera
    tenuifolia
    tenella
    viminea
    molissima
    nocturna
    decumbens
    odorata
    Romanzovii
    rosea-alba
    bifrons
    taraxifolia
Peas, Sweet Purple
    do. Scarlet
    sweet Striped
    do. White
    do. Yellow
    do. Black
    do. Painted Lady
    do. Top-knot

Peas, large Scarlet Tangier
    small do. do.
    Painted Lady do.
    Yellow-winged
    Red do.
    Lord Anson's
Persicaria, Red
    White
Poppy, Carnation
    Picotée
    New Fringed
    Double White
    Ranunculus or Dwarf
    Dwarf, Chinese
    Do. French
    in separate Colours
Prince's Feather, Red
    White
Roman Nettle
Rudbeckia amplexifolia
Scabious, Starry
Scabiosa prolifera
Snails
Sun-flower, tall
    dwarf
    extra double
Snap-dragon
Stock, Virginia, Red
    White
Strawberry Spinach
Silene Atocion
    colorata
    disticha
    major

Silene picta
    pendula
    rubella
    vespertina
Trifolium incarnatum
    cœruleum
Veronica, Species

Venus's Looking-glass
    do. White
    do. large Blue
    do. new Lilac
    do. Navel-wort
Vicia atropurpurea
Ximinesia encelioides

HALF HARDY,

Which should be sown in March, under Hand Glasses, or on a very moderate Hot-Bed, and transplanted into the Border in the middle of April or beginning of May.

Ageratum Mexicanum
    Odoratum
Anthemis Arabica
Argemone albiflora
    Mexicana
    Ochroleuca
Aster, Chinese, red
    ditto, purple
    ditto, white
    ditto, bonnet
    ditto, new early Dwf.
    superb red
    ditto, white
    ditto, blue
    fine quilled blue
    ditto, white
    ditto, red
    striped red
    ditto, purple
    new German varieties
    new Turkey red
Balm of Gilead
Blumenbachia insignis

Cacalia coccinea
Campanula gracilis
Cardiospermum Halicacabum
Carthamus tinctoria
Calceolaria pinnata
Calendrina speciosa
    grandiflora
Centaurea Americana
Cistus guttatus
Claytonia alsinoides
Clintonia elegans
Cyclobothria pulchella
    alba
Coix lachryma
Colutea, scarlet
Cuphea viscosissima
Datura ceratocaula
    Metel
    Tatula
Dendromecon rigidum
Dolichos Lablab
    purpureus

Galinsogea trilobata
Hollyhock, Chinese
Hibiscus Africanus
Hornemannia bicolor
Ipomœa Michauxii
    barbigera
    discolor
    hederacea
    hepaticæfolia
Jacobea, purple
    white
    fine double
Kaulfussia amelloides
Leonurus heterophyllus
    Nepalensis
Lopezia coronata
    racemosa
Marvel of Peru, red
    white
    yellow
    lg. tube or sweet scented
    gold-striped
    silver-striped
    red-striped
Marygold, orange African
    lemon ditto
    fine quilled
    dwarf French
    tall ditto
    striped ditto
    new early dwarf
    unique
Molucella lævis

Mesembryanthemum
    glabrum
Nicotiana angustifolia
    glutinosa
    alata
    quadrivalvis
    odorata
    new, scarlet
    rustica, or Maryland Tobacco
    Tabacum, or Virginia Tobacco
    undulata
Palma Christi major
    minor
Petunia Nyctaginiflora
Pink, Indian double
    ditto, broad-leaved
Salpiglossis picta
    straminea
    atropurpurea
    and others
Schizanthus pinnatus
    porrigens
    Hookeri
    Humilis
Stock, new Russian, in forty colours
    Prussian
    giant, ten week
    fine scarlet
    ditto purple
    ditto white

Stock, wall-leaved white
    ditto purple
    ditto scarlet
Sultan, sweet purple
    white
    yellow
Sycios angulata
Tagetes corymbosa
    tenuifolia
Trachymene cœrulea
Triteleia laxa
Xeranthemum, ann. white
    purple

Xeranthemum, lucidum
    (*Elychrysum*)
Zinnia elegans
    coccinea
    purpurea
    grandiflora
    fl. alba
    multiflora, red
        yellow
    revoluta
    tenuiflora
    verticillata
    grandiflora

TENDER ANNUALS,

Which require more than one Hot Bed to bring them to perfection, should be sown during the months of February and March.

Amaranthus, purple globe
    white ditto
    striped ditto
    bicolor
    tricolor
Balsam, fine double
    French varieties
Browallia elata, blue
    ditto, white
    demissa
Canna angustifolia
    coccinea
    indica
    lutea
Capsicum, long red
    yellow
    oxheart
    giant

Capsicum, cherry
    cayenne
    tomatoe
    bird pepper
Cleome spinosa
    pentaphylla
    rosea
Cockscomb, tall red
    fine dwarf
    branching
    pyramidal
    beautiful Chinese
    yellow and buff
Colutea, scarlet
Egg Plant, purple
    white
Hedysarum gyrans
Heliotropium Indicum

Heliotropium Peruvianum
Ice Plant
Ipomœa coccinea
    nil
    Quamoclit
Martynia annua

Martynia proboscidea
Mesembryanthemum cordifolium, or purple Ice Plant
Sensitive Plant
Stramonium, double purple
    ditto white

BIENNIALS & PERENNIALS, NOT HARDY.

Alstrœmeria peligrina
    ditto alba
    pulchella
    Simsii
    aurea, and others
Calceolaria suberecta

Calceolaria pendula
    mixed varieties
Gloxinia formosa
    speciosa
Thunbergia alata
and many others

HARDY BIENNIAL & PERENNIAL FLOWER SEEDS,

To be sown from the beginning of April to the middle of June.

Acanthus spinosa
Aconitum album
    Lycoctonum
    Napellus
    rubrum
    variegatum
Agrostemma Flos Jovis
Ammobium alatum
Anchusa sempervirens
    incarnata
Anemone coronaria
    narcissiflora
Anagallis grandiflora
    Monelli
Antirhinum bicolor
    pictum
    majus
    new scarlet
    new yellow
Alyssum saxatile

Aquilegia Canadensis
    alpina
Argemone grandiflora
Astragalus alopecuroides
    bœticus
    galegiformis
Astrantia, species
Auricula
Betonica grandiflora
Calceolaria corymbosa
    many varieties
Campion, Rose
Catananche cœrulea
Campanula azurea
    glomerata
    latifolia
    ditto alba
    percissiflora
    pyramidalis
    urticifolia

Campanula versicolor
Canterbury Bells, blue
               white
Carnation, fine double
       seedling various
Celsia cretica
      urticifolia
Chelone in varieties
Commelina cœlestis
        tuberosa
Cobæa scandens
Columbine
Cowslip, in colours
Convolvolus ciliatus
        Scammonia
Cynoglossum pictum
Cyclamen
Dahlia, fine double
Delphinium azureum
        elatum
        grandiflorum
        intermedium
Dianthus alpinus
     superbus
     latifolius
     velutinus
Digitalis alba
     aurea
     ferruginea
     grandiflora
     lutea
     ochroleuca
     orientalis
     rubra
Dracocephalum altaicum
        canescens
        peregrinum

Eccremocarpus scaber
Echium rubrum
     Sibericum
     plantaginium
     violaceum
Epilobium spicatum
Eryngium alpinum
     amethystinum
     planum
Eschscholtzia californica
     crocea
Francoa appendiculata
Ferula tingitana
Flax, perennial
     Siberian
Fraxinella, red
     white
Galardia aristata
     bicolor
Galega officinalis, white
        blue
Gaura mutabilis
Geranium, from choice sorts
     Wallachianum
Gentiana acaulis
     asclepiadea
     cruciata
     saponaria
Geum album
     coccineum
     Quellion
Glaucium fulvum
     luteum
Hesperis matronalis
Heart's-ease
Hemerocallis cœrulea
Hieracium aureum

Hieracium maculatum
Hollyhock, double mixed
   in separate colours
      Antwerp
Honesty, or Lunaria
Honeysuckle, red French
      white ditto
Horn Poppy
Iris Florentina
   Siberica
   Susiana
   Xiphium
   Xiphioides
Ketmia versicaria
Lophospermum scandens
Larkspur, perennial
Lespedezia capitata
Liatris spicata
     scariosa
     elegans
Linaria alpina
     genistifolio
     purpurea
Lobelia cardinalis
    syphilitica
    bignonifolia
Lupinus lepidus
     mexicanus
     ornatus
     perennis
     plumosus
     polyphillus
     ditto albus
     tomentosus
     nanus
Lychnis fulgens
    scarlet
    white

Lythrum erubescens
Lathyrus latifolius
    heterophyllus
Linum, large new
Malva arborea
   moschata
Maurandia Barclayana
Mimulus guttatus
   luteus
   moschatus
   pictus
   rivularis
   roseus
Monkshood, various
Monarda punctata
   oblongata
Myagrum, species
Medicago arborea
Nierembergia, Phœnicia
   rosea
Oenothera biennis
   grandiflora
   fruticosa
   pumila
   spectabilis
Ononis hircina
   rotundifolia
Orobus niger
   vernus
   coccinea
   angustifolius
Papaver, bracteatum
   orientale
Peas, Everlasting
Penstemon atrorubens
   atropurpureum
   angustifolium
   Richardsonii

Penstemon diffusum
    digitale
    hybridum
    Lyoni
    ovatum
    pulchellum
Peony, in sorts
Phytolacca decandra
Pink, pheasant-eyed
    mountain
    Spanish
Polemonium, white
    blue
Polyanthus
Prunella grandiflora
Potentilla atrosanguinea
    formosa
    hirta
Primula prænitens
    prænitens alba
Psoralea grandiflora
Pulsatilla alpina
    vulgaris
Pulmonaria Siberica
Ranunculus
    aconitifolius
Rhodiola rosea
Rocket, sweet
Ruellia australis
Rudbeckia hirta
    laciniata
    purpurea
Scabiosa alpina
    caucasica
    moschata or sweet
    nana
Serratula alata
Spiræa Aruncus

Sisyrinchium Nuttalli
Statice latifolia
    tartarica
Stevia purpurea
Stipa pinnata
Stock, scarlet Brompton
    white Brompton
    purple ditto
    Twickenham
    scarlet Queen
    white ditto
    purple ditto
Tagetes lucida
Thistle, Globe
Trollius asiaticus
    europeus
    giganteus
Valerian, scarlet
    white garden
    Greek, white & blue
Veratrum nigrum
    viride
Verbena aubletia
    Lambertii
Verbascum Blattaria
    biennis
    formosum
    Ibericum
    Myconi
    pyramidatum
Veronica spicata
Vesicaria utriculata
Viola odorata
Wallflower, blood red
    yellow
William, Sweet, scarlet
    do. auricula-flowered

Price 25 cts.

GEO. C. THORBURN'S

CATALOGUE

OF

KITCHEN GARDEN,

HERB, FLOWER, TREE AND GRASS

SEEDS,

BULBOUS FLOWER ROOTS,

GREENHOUSE PLANTS,

GARDENING, AGRICULTURAL, AND BOTANICAL

BOOKS,

GARDENING TOOLS,

ETC. ETC.

CATALOGUE

OF

KITCHEN GARDEN,

HERB, FLOWER, TREE, AND GRASS

SEEDS;

BULBOUS FLOWER ROOTS;

GREENHOUSE PLANTS;

GARDENING, AGRICULTURAL, AND BOTANICAL

BOOKS;

GARDEN TOOLS, ETC.

WITH

AN APPENDIX,

Containing a variety of useful Agricultural information.

SOLD BY

GEO. C. THORBURN,

SEEDSMAN AND FLORIST,

No. 11 John Street, near Broadway.

"As Gardening has been the inclination of kings, and the choice of philosophers, so it has been the favourite of public and private men; a pleasure of the greatest, and the care of the meanest: and, indeed, an employment and profession for which no man is too high nor too low.—TEMPLE.

NEW-YORK:

GEORGE P. SCOTT AND CO. PRINTERS.

1838.

CONTENTS.

Seeds of Esculent Vegetables........5	Pæonies........................53
Directions for the cultivation of Vegetables........................13	Miscellaneous Sorts............54
Pot and Sweet Herbs..............26	Directions for the Management of Bulbous and other Flower Roots, and for the care of Plants in Rooms.55
Medicinal Herb Seeds.............26	To Destroy the Curculio.............59
Esculent Roots and Plants.........27	Greenhouse Plants..................60
Annual Flower Seeds..............28	Roses..............................69
Biennial and Perennial Flower Seeds.33	Geraniums..........................71
Observations on the Flower Garden..38	Dahlias............................73
Native American Tree, Shrub, and Plant Seeds........................41	Management of Greenhouse Plants..74
Agricultural or Farming Seeds......44	Garden Tools.......................77
Dutch Bulbous Flower Roots........46	Agricultural, Botanical, and Gardening Books........................79
Double Hyacinths..............46	Miscellaneous and Fancy Articles...86
Single Hyacinths..............48	APPENDIX.
Tulips........................49	Lucerne......................88
Amaryllis.....................51	Orchard Grass................89
Crown Imperials...............51	Baden Corn...................91
Crocus........................51	Italian Spring Wheat.........92
Fritillaries...................51	Eley's New Gigantic Winter Wheat......................94
Gladiolus, or Sword Lily.......51	White Mulberry Tree..........94
Iris, or Flower De Luce........52	Hawthorn Hedges..............95
Ixias.........................52	Rohan Potato.................97
Lachenalias...................52	Bene Plant...................98
Lilies........................52	Dutton Corn..................98
Martagons, or Turk's Cap Lilies.52	Toronto......................99
Polyanthus Narcissus..........52	Estimate of Seeds for an acre....100
Double Narcissus..............53	
Single Narcissus..............53	

Prove seeds by sowing a few in a small pot or box of light loose earth, and placing it in a warm room, exposed to the sun: keep the earth moist, and if the seed is good, it will vegetate in a reasonable time, except peas and beans, which should be tried by sowing a few in the open ground.

The *bug-holes* frequently seen in peas, are not occasioned by age, but are the work of an insect, which deposites its eggs in the flower, and matures with the pea; when it eats its way out at the side, leaving the *eye* of the pea uninjured; this of course does not destroy its vegetating power.

If the seeds do not grow after being tried as above, they may be returned, when they will be replaced with other seeds, or the money refunded.

Celery, spinage, onion, and, indeed, almost every other kind of seed, will not vegetate, except the ground is rolled after the seed is sown, or except a heavy rain falls to beat the ground, which answers the same purpose. Where there is no roller on the premises, the following may answer for a substitute.

After the seed is sown and the ground well raked, take a board, (or boards,) the whole length of the bed—lay them flat on the ground, beginning at one edge of the bed—walk the whole length of the board—this will press the soil on the seed—then shift the board till you have thus gone over the whole bed; and, in dry weather, cover your seed-beds for forty-eight hours, with boards laid flat on the soil, and the seeds will come up almost immediately. If no boards are at hand, tread in the seed with your feet, or strike on the beds with your spade or shovel.

Those who wish to purchase to sell again, can be supplied on very advantageous terms. A list of the wholesale prices, on a letter sheet, can be obtained on application; or the seeds neatly put up in six-cent papers, labelled and sealed ready for retail; from which we make a very liberal discount, and warrant every sort to be of the very first quality.

PREFACE.

In arranging the materials for this Catalogue, the subscriber has thought it most convenient for reference to keep the Vegetables together, and to place the directions for their cultivation immediately afterwards; and in the same manner with the Flowers, and the Greenhouse department.

His collection of *bulbous flowers* has been greatly enriched by many new and fine varieties from Holland; and among his *Dahlias* will be found the latest and most splendid kinds originated in England, whose gardeners, it is well known, are celebrated for their skill in floriculture. The additions to his *camellias* and *Geraniums*, are of the most rare and magnificent description. He has procured many new and beautiful *Annuals* and *Biennials*; and his collection of *Kitchen Garden Seeds* contains every useful kind of vegetable cultivated either in private gardens, or for the market. The whole of which he takes this opportunity of recommending to his friends and the public, with the assurance that they may always rely on being served with the best and most genuine articles.

The following sensible observations are so well expressed on the beneficial effects of flower culture, that he takes great pleasure in giving them a place here.

"The interest which flowers have excited in the breast of man, from the earliest ages to the present day, has never been confined to any particular class of society, or quarter of the globe. Nature seems to have distributed them over the whole world, to serve as a medicine to the mind, to give cheerfulness to the earth, and to furnish agreeable sensations to its inhabitants.

"The savage of the forest, in the joy of his heart, binds his brow with the native flowers of the woods, whilst a taste for their cultivation increases in every country, in proportion as the blessings of civilization extend.

"From the humblest cottage enclosure, to the most extensive park and grounds, nothing more conspicuously bespeaks the good taste of the possessor, than a well cultivated flower garden; and it may very generally be remarked, that when we behold an humble tenement surrounded with ornamental plants, the possessor is a man of correct habits, and possesses domestic comforts; whilst, on the contrary, a neglected, weed-grown garden, or its total absence, marks the indolence and unhappy state of those who have been thus neglectful of Flora's favours.

"Of all luxurious indulgences, that of flowers is the most innocent. It is productive not only of rational gratifications, but of many advantages of permanent character. Love for a garden has a powerful influence in attracting men to their homes; and, on this account, every encouragement given to increase a taste for ornamental gardening, is additional security for domestic comfort and happiness. It is likewise a recreation which conduces materially to health, promotes civilization, and softens the manners and tempers of men. It creates a love for the study of nature, which leads to a contemplation of the mysterious wonders that are displayed in the vegetable world around us, and which cannot be

PREFACE.

investigated without inclining the mind towards a just estimate of religion, and a knowledge of the narrow limits of our intelligence, when compared with the incomprehensible power of the Creator.

"Flowers are, of all embellishments, the most beautiful, and of all created beings, man alone seems capable of deriving any enjoyment from them. The love for them commences with infancy, remains the delight of youth, increases with our years, and becomes the quiet amusement of our declining days. The infant can no sooner walk, than its first employment is to plant a flower in the earth, removing it ten times in an hour to wherever the sun seems to shine most favourable. The school-boy, in the care of his little plot of ground, is relieved of his studies, and loses the anxious thought of the home he has left. In manhood, our attention is generally demanded by more active duties, or more imperious, and perhaps less innocent occupations; but as age obliges us to retire from public life, the love of flowers, and the delight of a garden, return to soothe the latter period of our life.

"To most persons, gardening affords delight as an easy and agreeable occupation; and the flowers they so fondly rear, are cherished from the gratification they afford to the organs of sight and sense; but to the close observer of nature, and the botanist, beauties are unfolded and wonders displayed, that cannot be detected by the careless attention bestowed upon them by the multitude.

"In their growth, from the first tender shoots which rise from the earth, through all the changes which they undergo, to the period of their utmost perfection, he beholds the wonderful works of creative power; he views the bud as it swells, and looks into the expanded blossom, delights in its rich tints, and fragrant smell, but above all, he feels a charm in contemplating movements and regulations, before which all the combined ingenuity of man dwindles into nothingness."

GEO. C. THORBURN.

NEW-YORK, 1833.

Mushroom Spawn	Agaricus campestris
Potato Onion	Allium var.
Top or Tree Onion	———— var.
Tarragon, or Astragon	Artemisia dracunculus
Asparagus	Asparagus officinalis.

[75 cents per hundred, $8 per thousand.]

Horse Radish	Cochlearia armoracia
Caroline Sweet Potato Slips	Convolvulus batatas
Sea Kale	Crambe maritima
Jerusalem Artichoke	Helianthus tuberosus
Hop	Humulus lupulus
Undulated Rhubarb	Rheum undulatum
Early Nonpareil Potatoes, (and other fine sorts)	Solanum tuberosum
English White Kidney do.	———— var.
Lemon Thyme	Thymus serpyllum.

ANNUAL FLOWER SEEDS.

SIX CENTS PER PAPER.

Those marked [c] are climbing plants.

Flos Adonis, or Pheasant's Eye	Adonis miniata
Mexican Blue Ageratum	Ageratum mexicana
Sweet Scented do.	———— odoratum
Sweet Alyssum	Alyssum maritimum
Love Lies Bleeding	Amaranthus caudatus
Straw colored do.	———— var. lutea
Prince's Feather	———— hypocondriacus
New Beautiful Crimson Amaranthus	———— hypocondriacus nova
Three colored Amaranthus	———— tricolor
Blue Pimpernell	Anagallis indica
Great flowering Argemone	Argemone grandiflora
China Aster, Early Dwarf	Aster sinensis
——— White	——— fl. albo
——— Anemone flowered	——— anemoniflora
——— Purple	——— fl. purpureo
——— Maiden's Blush, or Rose	——— fl. incarnata
——— Lilac	——— fl. obscuro
——— Superb Quilled	——— fl. superba
——— Bonnet	——— var. nova

Annual Flower Seeds.

China Aster, Red Striped	Aster rub. variegato
—— Purple Striped	—— pur. variegato
—— Red	—— fl. rubro
—— Blue	—— cerulea
—— New Crimson Turkey	—— fine var.
Alkekengi, or Kite Flower	Atropha physaloides
Animated Oats	Avena sensitiva
Strawberry Spinach	Blitum capitatum
Browallia, or Amethyst	Browallia elata
White do.	—— fl. albo
Quaking Grass, or Ladies' Tresses	Brixa maxima
Scarlet Cacalia, or Tassel flower	Cacalia coccinea
Marigold Starry	Calendula stellata
—— Great Cape	—— hybrida
'Balloon Vine, or love in a puff	Cardiospermum halicacabum
Lilac Venus Looking Glass	Campanula speculum purpureum
Venus' Looking Glass	Campanula speculum
Safflower, or Saffron	Carthamus tinctorius
Sensitive Cassia	Cassia nictitans
Long rayed American Centaurea	Centaurea Americana
Blue Bottle, Great	—— cyanus major
—— Small	—— —— minor
Sweet Sultan, Purple	—— moschata
—— Yellow	—— suaveolens
Blessed Thistle	—— benedicta
Cockscomb, Crimson Velvet	Celosia cristata
—— Yellow	—— var. lutea
Great Honeywort	Cerinthe major
Belvidere, or Summer Cypress	Chenopodium scoparia
Giant ten week Stockgillyflower (mixt colors)	Cheiranthus annuus
Virginian Stockgilliflower	—— maritimus
Chrysanthemum, White	Chrysanthemum coronarium
—— Yellow	—— fl. lutea
—— Tricolored	—— tricolor
Beautiful Clarkea	Clarkea pulchella
Elegant Clarkea	Clarkea elegans
Job's Tears	Coix lachryma Jobi
Great Flowered Collinsia	Collinsia grandiflora
Blue Commelina	Commelina coelestes
Dwarf Convolvulus	Convolvulus minor
'Morning Glory, Yellow	—— fl. lutea
'—— Azure	—— nil
'—— Dark Blue	—— major

3

ᶜMorning Glory, Rose colored	Convolvulus, fl. rosea
ᶜ———— Striped	———— fl. striata
Elegant Coreopsis	Coreopsis elegans
Late Cosmos	Cosmos bipinnatus
Venus Navelwort	Cotydelon malocophyllum
Hawkweed, Golden	Crepis barbata
———— Red	———— rubra
ᶜGourd, Two colored Lemon Shaped (*beautiful*)	Cucurbita bicolor
ᶜ———— Orange	———— aurantia
———— Large Bottle	———— lagenaria
———— Club-fruited	———— clavata
Orleans Vine	Cucumus chate
ᶜSnake Melon	———— melo anguinus
ᶜPomegranate, or Sweet Scented Melon	———— odoratissimus
White do. do. do.	———— fl. albo
White Sweet Sultan	Cyanus moschatus alba
Larkspur, Branching	Delphinium consolida
———— Double Rose	———— fl. roseo
———— Double Dwarf Rocket	———— var.
———— Dwarf Neapolitan	———— var.
Pink, Profuse Flowering	Dianthus prolifera
———— Annual China	———— annuus
ᶜHyacinth Bean, Purple	Dolichos lablab
ᶜ———— White	———— var. albo
Variegated Euphorbia	Euphorbia variegata
Musk Geranium	Geranium moschatum
Horned Poppy	Glaucium luteum
Globe Amaranthus, Purple	Gomphrena globosa
———— White	———— fl. albo
Cotton Plant	Gossipyum herbaceum
Azure Blue Gilia	Gilia capitata
White Gilia	Gilia capitata alba
Sunflower, Mexican	Helenium mexicanum
———— Tall	Helianthus annuus
———— Dwarf	———— v. nanus
African Hibiscus	Hibiscus africanus
Bladder Catmia	———— trionum
Candytuft, White	Iberis amara
———— Purple	———— umbellata
———— New Fine Purple	———— ———— var. speciosa
Double Balsamine, Fine Mixed	Impatiens balsamina
———— Rose colored	———— fl. roseo
Balsam, Variegated	———— variegata
———— Fire colored	———— coccinea
———— Purple	———— purpurea

Annual Flower Seeds.

Balsam, Pure White	Impatiens alba
——— Crimson	——— rubro pleno
Scarlet Morning Glory	Ipomœa coccinea
ᶜIpomœa Starry	——— lacunosa
——— hepatica leaved	——— hepaticafolia
Cypress Vine, Crimson	——— quamoclit
——— White	——— —— fl. albo
ᶜSweet Peas, Painted Lady Topknot	Lathyrus odoratus flore carneo
ˢ——— Yellow	——— aphaca
ᶜ——— White	——— var. albo
ᶜ——— Black	——— fl. obscuro
ᶜ——— Purple	——— fl. purpureo
ᶜ——— Scarlet	——— fl. roseo
ᶜ——— Striped	——— sativus
ᶜLord Anson's Peas	——— fl. striata
ᶜTangier Crimson Peas	——— tingitanus
ᶜWinged Peas	Lotus tetragonolobus
Red Lavatera	Lavatera trimestris
White Lavatera	——— trimestris alba
Lupins, White	Lupinus albus
——— Yellow	——— luteus
——— Large Blue	——— pilosus
——— Dutch do.	——— hirsutus
——— Small do.	——— varius
——— Rose	——— fl. roseo
Splendid Yellow Madea	Madea splendens
Scarlet Flowered Malope	Malope trifida
Curled Standing Mallow	Malva crispa
Cuckold's Horn	Martynia diandria
Proboscis Capsuled Martynia	——— proboscidea
Caterpillars	Medicago circinnata
Hedgehogs	——— intertexta
Snails	——— scutellata [num
Ice Plant	Mesembryanthemum crystalli-
Red, or Dew Plant	——— glabrum
Blue Veitch	Melilotus cerulea
Sensitive Plant	Mimosa sensitiva
Marvel of Peru	Mirabilis jalapa
Sweet Scented do.	——— longiflora
Squirting Cucumber	Momordica elaterium
ᶜBalsam Apple	——— balsamina
ᶜBalsam Pear	——— lagernia
Forget-me-not	Myosotis arvensis
Tobacco, Yellow Virginia	Nicotiana rustica
——— Scarlet flowering Havana	——— tabacum
——— Sweet Scented	——— odorata

Devil in a Bush, or Love in a Mist	Nigella damascena
Dwarf Love in a Mist	——— damascena nana
Trailing Nolana	Nolana prostrata
Touch-me-not	Noli mi tangere
Evening Primrose, Yellow	Œnothera grandiflora
——— Night smelling	——— nocturna
——— Dwarf Blue	——— tenella
——— White	——— tetraptera
——— Purple and White	——— lindleyana
——— Blue	——— romanzovii
——— Rose colored	——— roseo
Primrose, Purple	——— purpurea
Poppy, White Officinal	Papaver somniferum
——— Double Carnation	——— fl. pleno
——— Corn or Rose	——— rhæas
——— Ranunculus	——— var.
——— Marginate	——— marginata
——— Dwarf Chinese	——— sinensis
——— Lap-dog	——— bichon
——— Dwarf Dutch	——— var.
ᶜScarlet Pentapetes	Pentapates phœnicea
ᶜBean, Scarlet Flowering	Phaseolus multiflorus
——— Dwarf Flowering	——— superba
Red Persicaria	Polygonum orientale
Mignionette	Reseda odorata
ᶜWhite Egg Plant	Solanum melongena
Starry Scabious	Scabiosa stellata
Dwarf Purple Scabious	——— atropurpurea nana
Schizanthus, Wing Leaved	Schizanthus pinnatus
——— Showy	——— porrigens
Tangier, or Poppy-leaved Viper's Grass	Scorzonera tingitana
Jacobea, Purple (*double*)	Senecio elegans
——— White (*double*)	——— fl. albo
Catchfly	Silene armeria
Dwarf Catchfly	——— rubella
Evening Catchfly	——— vespertina
Painted Catchfly	——— picta
Vanilla Scented Stevia	Stevia serrata
Feather Grass	Stipa pinnata
Marigold, African	Tagetes erecta
——— Orange quilled	——— fl. teretibus
——— French	——— patula
——— Ranunculus	——— var.
New Early Dwarf	——— nova
Trefoil, Crimson	Trifolium incarnatum

Trefoil, Sweet Scented	Trifolium odorata
Nasturtium, superb large crimson	Trapæolum atrosanguinea
——— Dwarf	——— v. nana
Heartsease, or Pansy	Viola tricolor
White Immortal Flower (*fine*)	Xeranthemum bracteatum alba
Eternal Flower, Golden	——————— lucidum
——— Purple	————— annum
Mexican Ximenisia	Ximenisia enceloides
Great Flowering Red Zinnia	Zinnia multiflora grandiflora
Zinnia Yellow	——— pauciflora
——— Violet colored	——— elegans

BIENNIAL AND PERENNIAL FLOWER SEEDS.

Graines de fleurs bisannuelles et vivaces.

SIX CENTS PER PAPER.

Those marked [d] are delicate, and require to be housed in winter.
Those marked [c] are climbing plants.

Monk's hood	Aconitum napellus
Rose campion	Agrostemma coronaria
Pink Rose Campion, with white centre	Agrostemma rosea albo
White Campion, beautiful	Agrostemma albo
Hollyhock, black Antwerp	Althea fl. nigra
——— double yellow	——— flava pl.
——— double china	——— sinensis
Golden alyssum	Alyssum saxatile
Anemone, or wind flower	Anemone coronaria
Musk scented genarium, or pasque flower	Anemone pulsatilla
Beautiful Variegated Crimson and White Snapdragon	Antirrhinum majus, variegata
Snapdragon, scarlet	——— majus
——— two colored	——— bicolor
Double columbine	Aquilegia vulgaris
Swallow-wort, orange	Asclepias tuberosa
——— starry	——— incarnata
cScarlet trumpet flower	Bignonia radicans
Canterbury bell, blue	Campanula medium
——— white	——— fl. albo

CATALOGUE

OF

A CHOICE COLLECTION

OF

FLOWER SEEDS,

FOR

1845,

COMPRISING, IN ALL, UPWARDS OF

FOUR HUNDRED SPECIES AND VARIETIES,

INCLUDING

SOME SPLENDID ASSORTMENTS

OF

German Asters, Balsams, Ten-Week-Stocks, Larkspurs, Hollyhocks, Candytufts, Poppies, Zinnias, &c.

AMONG THE SEEDS ARE MANY NEW, RARE AND SPLENDID VARIETIES, WELL WORTHY THE ATTENTION OF FLORISTS AND AMATEUR CULTIVATORS.

CULTIVATED, IMPORTED AND FOR SALE
BY

JOSEPH BRECK & CO.

SEEDSMEN AND NURSERYMEN,

Nos. 51 & 52 North Market Street, (up Stairs,)

BOSTON.

EXPLANATIONS.

The number under which each species or variety of Seed is sold—the Scientific Name—the Common Name—the Period of Duration of the Plant—Color of the Flowers—Height of the Plant—Period of Blooming—and Price per Packet, are all given in the Catalogue, as follows:—

1st Column.—Number under which each variety is sold—and under which orders are executed, the detail of the names being unnecessary.

2d Column.—Scientific or common name of the plants, agreeably to the remarks at the head of each Alphabetic List.

3d Column.—The same.

4th Column.—Hardiness and duration of each plant, viz:—h, *hardy;* hh, *half hardy;* t, *tender;* F, *frame;* G, *green-house;* a, *annual* (1 year;) b, *biennial* (lasts 2 years;) p, *perennial* (last many years) They are thus applied in the Catalogue : —ha, *hardy annual;* hha, *half hardy annual;* ta, *tender annual;* hb, *hardy biennial;* hhb, *half hardy biennial;* hp, *hardy perennial;* hhp, *half hardy perennial,* &c.

5th Column.—Color of the flower. The abbreviations are as follows :—var., *various;* sc., *scarlet;* cr, *crimson;* pur., *purple;* str. *striped;* yel. *yellow;* va., *variegated;* or., *orange;* b. & w., *blue and white,* &c.

6th Column.—Usual height in feet the plants generally attain under good cultivation.

7th Column.—Usual months of flowering.

8th Column.—Price of the seeds per single packet.

A star, *, added to the letters of the fourth column, denotes that the biennial and perennial plants flower the first year as well as the second.

A double star, **, following the scientific name, signifies that the plants are climbing, and suitable for an arbor, or trellis work.

TIME OF SOWING.—Hardy annuals from April to June, and many of the kinds in the autumn. Half hardy annuals in May, or earlier, in a *green-house* or a *hot bed.* Tender annuals in a *hot bed,* and transplanted to the border in June. Biennials and perennials from April to July.

ASSORTMENTS
OF
SPLENDID FLOWER SEEDS,

Comprising superb Double German Asters, Balsams, Larkspurs, &c., raised by ourselves, and warranted to be of the most splendid description.

No.		Price.
1. 10 varieties of superb Double German Asters	- - - - -	50
2. 12 varieties of beautiful German Ten Week Stocks	- - - - -	62½
3. 4 varieties do. do. do.	- - - - -	25
4. 6 varieties of Sweet Peas	- - - - -	37½
5. 4 varieties of superb Double Balsams	- - - - -	25
6. 6 varieties Lupin	- - - - -	37½
7. 4 varieties of fine Double Poppies	- - - - -	25
8. 6 varieties splendid Lupin	- - - - -	62½
9. 4 varieties of Goodetias, handsomest colors	- - - - -	25
10. 12 varieties of splendid Larkspurs	- - - - -	62½
11. 10 varieties of fine Double Hollyhocks	- - - - -	50
12. 4 varieties of fine Candytuft	- - - - -	25
13. 4 fine Schizanthuses	- - - - -	25
14. 4 varieties of fine Zinnias	- - - - -	25
15. 20 varieties of Flower Seeds assorted kinds	- - - - -	1 00
16. 50 varieties do. do. do. do.	- - - - -	2 00

CATALOGUE.

The following plants, being best known by their common, or English names, they are thus arranged in alphabetical order—and the scientific name given in the third column. A collection may be ordered, by merely giving the numbers and date of the Catalogue.

Number.	Common Name.	Scientific Name.	Duration of Plant.	Color of Flower.	Height in feet.	Period of Flowering.	Price.
1	Adonis, Flos	Adonis æstivalis	ha	scarlet	1	July, Aug.	6
2	Animated Oats	Avena sensitiva	ha	white	1	do	6
3	Aster, Chinese, mixed	Aster sinensis, var.	hha	var.	2	Aug. Oct.	6
4	German, mixed	sp. and var.	"	var.	2	do	6
5	blue	cæruleus	"	blue	2	do	6
6	light blue	læte-cæruleus	"	pale blue	2	do	6
7	red	atrorubens	"	red	2	do	6
8	rose	roseus	"	rose	2	do	6
9	white	alba	"	white	2	do	6
10	ash gray	var.	"	ash	2	do	6
11	red and white	rubro albus	"	red & w.	2	do	6
12	blue and white	cæruleo albus	"	b. and w.	2	do	6
13	turkey	turcicus	"	red	2	do	6
14	early dwarf	nanus	"	var.	2	July, Sept.	6
15	Auricula, fine mixed	Primula auricula	hhp	var.	½	May, June	12
16	Anagallis, mixed	Anagallis, sp. and var.	ha	var.	1	July, Aug.	6
17	Balsam, fine mixed	Balsamina hortensis	ta	var.	2	July, Sept.	6
18	striped	striata	"	striped	2	do	6
19	rose	rosea	"	rose	2	do	6
20	scarlet	coccinea	"	scarlet	2	do	6
21	white	alba	"	white	2	do	6
22	ruby	var.	"	ruby	2	do	6
23	purple	purpureus	"	purple	2	do	6
24	new mottled	punctata var.	"	red & w.	2	do	12
25	crimson spot'd	punctata var.	"	c. and w.	2	do	12
26	scarlet spotted	punctata var.	"	sc. & w.	2	do	12
27	purple spotted	punctata var.	"	pur. & w	2	do	12
28	Chinese rose	sinensis rosea	"	rose	2	do	12
29	Cape Marygold	Calendula pluvialis	ha	w. and p.	2	do	6
30	Candytuft, fragrant	Iberis odorata	"	white	1	June, Aug.	6
31	purple	umbellata	"	purple	1	do	6
32	Normandy	major	"	purple	1	do	6
33	new crimson	phœnicea	"	crimson	1	do	6
34	superb rocket	coronaria	"	white	1	do	6
35	white	amara	"	white	1	do	6
36	Calliopis, golden	Calliopsis tinctoria	"	y. & pur.	2	June, Sept.	6
37	Drummond's	Drummondii	"	yellow	2	do	6
38	dark red	atrosanguinea	"	y. and cr.	2	do	6
39	Calceolaria, fine mixed	Calceolaria var.	hp	var.	2	May, Aug.	12
40	Carnation, mixed	Dianthus caryophylloi-	"	var.	2	June, Aug.	12
41	extra fine	var. [des	"	var.	2	do	25
42	Picotee	var.	"	var.	2	do	6

JOSEPH BRECK AND CO.'S CATALOGUE OF FLOWER SEEDS.

Number.	Common Name.	Scientific Name.	Duration of Plant.	Color of Flower.	Height in feet.	Period of Flowering.	Price.
43	Carnation, extra fine	Dianthus var.	hp	var.	2	June, Aug.	25
44	Catchfly, Lobels	Silene armeria	ha	red	1	do	6
45	large cluster'd	compacta	"	rose	1½	do	6
46	white	alba	"	white	1	do	6
47	Canterbury Bell, blue	Campanula medium	hb	blue	2	June, Sept.	6
48	white	alba	"	white	2	do	6
49	Coxcomb, crimson	Celosia cristata	ta	cr.	1½	July, Sept.	6
50	yellow	lutea	"	yel.	1½	do	6
51	Columbine, mixed	Aquilegia vulgaris	hp	var.	2	May, July	6
52	Siberian	Siberica	"	blue	1	do	6
53	Chrysanthemum, yell.	Chrysanthemum coro-	ha	yel.	2½	July, Aug.	6
54	white	alba [narium	"	white	2½	do	6
55	tricolored	carinatum	"	3 col.	1	do	6
56	new golden	luteum	"	yel.	1	do	6
57	Chinese Primrose	Primula prænitens	hhp	purp.	¼	July, Dec.	25
58	white	alba	"	white	½	do	25
59	Cypress vine,	Ipomæa Quamoclit	ta	cr.	10	July, Sep.	6
60	Convolvulus, dwarf	Convolvulus minor	ha	bl. & wh.	1½	July, Aug.	6
61	Cacalia, scarlet	Cacalia coccinea	"	sc.	1	do	6
62	new golden	lutea	"	yel.	1	do	6
63	Dahlia, finest mixed	Dahlia, var.	tp	var.	5	July, Oct.	12
64	Evening Primrose	Œnothora grandiflora	ha	yel.	3	July, Sept.	6
65	Egg Plant, purple	Solanum melongena	ta	purp.	2	do	6
66	white	var.	"	white	2	do	6
67	Eternal Flower, golden	Elichrysum bracteatum	hha	purp.	2	do	6
68	new white	album	"	white	2	do	6
69	common purple	Xeranthemum annu-	"	yel.	1	do	6
70	common white	album [um	"	white	1	do	6
71	Forget-me-not	Myosotus arvensis	ha	blue	½	July, Aug.	12
72	Fraxinella, purple	Dictamnus fraxinella	hp	red	2	June, Aug.	6
73	white	alba	"	white	2	do	6
74	Foxglove, purple	Digitalis purpurea	"	purp.	2½	July, Sept.	6
75	white	alba	"	white	2½	do	6
76	Globe Amaranthus	Gomphrena glohosa	ta	cr.	2	do	6
77	white	alba	"	white	2	do	6
78	striped	striata	"	str.	2	do	6
79	Gourd, bottle**	Cucurbita lagenaria	"	white	10	do	6
80	striped pear**	ovifera var.	"	yel.	10	do	6
81	orange**	aurantia	"	yel.	10	do	6
82	mandrake**	var.	"	yel.	10	do	6
83	Geranium (or Pelargo-nium) extra fine	Pelargonium sp. & var.	gp	var.	2	April, June	25
84	Hawkweed, golden	Crepis barbata	ha	yel.	1	July, Aug.	6
85	Heartsease, mixed	Viola tricolor	"	var.	¾	May, Sept.	6
86	fine mixed	grandiflora	hb*	var.	¾	do	12
87	extra, from very fine named flowers	var.	"	var.	¾	do	25
88	Hibiscus, African	Hibiscus Africanus	ha	yel. & br.	1½	July, Sept.	6
89	swamp	palustris	hp	rose	4	June, Aug.	6
90	Hyacinth beans, pur-	Lablab vulgaris	hha	purp.	10	July, Sept.	6
91	white** [ple**	alba	"	white	10	do	6
92	Honesty, or Satinflower	Lunaria biennis	hb	blue	1½	July, Aug.	6
93	Hollyhocks, mixed	Althea rosea	hp	var.	5	June, Sept.	6
94	black	nigra	"	black	5	do	6
95	yellow	lutea	"	yel.	5	do	6
96	rose	rosea	"	rose	5	do	6
97	pink	rubella	"	pink	5	do	6
98	white	alba	"	white	5	do	6
99	sulphur	sulphurea	"	sulph.	5	do	6
100	crimson	coccinea	"	cr.	5	do	6
101	mottled	var.	"	mot.	5	do	6
102	Ice plant	Mesembryanthemum,	ta	white	½	do	6
103	Indian Shot	Canna indica [sp.	"	sc.	3	July, Aug.	6
104	Jacobæa, purple	Senecio elegans	hha	purp.	1	July, Sept.	6
105	white	alba	"	white	1	do	6
106	double purple	purpurea	"	purp.	1	do	6

JOSEPH BRECK AND CO.'S CATALOGUE OF FLOWER SEEDS.

Number.	Common Name.	Scientific Name.	Duration of Plant.	Color of Flower.	Height in feet.	Period of Flowering.	Price.
107	Larkspur, dwarf rocket	Delphinium ajacis	ha	var.	1½	June, Sept.	6
108	blue	cæruleum	"	blue	1½	do	6
109	porcelain	var.	"	p. blue	1½	do	6
110	white	album	"	white	1½	do	6
111	rose	rosea	"	rose	½	do	6
112	branching mixed	Consolida var.	"	var.	2½	July, Aug.	6
113	blue	cæruleum	"	blue	2½	do	6
114	red	roseum	"	rose	2½	do	6
115	Bee	elatum	hp	blue	4	June, Aug.	6
116	Chinese	chinensis	"	blue	3	do	6
117	large flowered	grandiflora	"	blue	3	do	6
118	Lavatera, red	Lavatera trimestris	ha	rose	2	July, Sept.	6
119	white	alba	"	white	2	do	6
120	Love Lies Bleeding	Amaranthus caudatus	"	red	2	do	6
121	Lupins, mixed	Lupinus sp. and var.	"	var.	2	do	6
122	large blue	hirsutus	"	blue	2	do	6
123	rose	pilosus	"	rose	2	do	6
124	small blue	angustifolius	"	blue	2	do	6
125	yellow	luteus	"	yel.	2	do	6
126	white	albus	"	white	2	do	6
127	Marygold, French	Tagetes patula, pl. var.	hha	yel. & br.	2	do	6
128	African	erecta	"	yel.	2½	do	6
129	Marvel of Peru, mixed	Mirabilis dichotoma	ha	var.	3	July, Oct.	6
130	Mignonette, sweet	Reseda odorata	"	white	1	June, Sept.	6
131	tree, or branching	var.	hb*	white	1	do	6
132	Nasturtium, common**	Tropæolum majus	ha	yel.	6	do	6
133	dark red**	atrosanguineum	"	cr.	6	July, Sept.	6
134	spotted**	Shillingii	"	spot.	6	do	6
135	Peas, sweet, mixed**	Lathyrus odoratus var.	"	var.	5	do	6
136	striped**	striatus	"	str.	5	Aug. Sept.	6
137	Painted Lady	pictus	"	var.	5	do	6
138	white**	albus	"	white	5	do	6
139	black**	nigrus	"	black	5	do	6
140	scarlet**	coccineus	"	sc.	5	do	6
141	purple**	purpureus	"	purp.	5	do	6
142	Tangier**	tingitanus	"	sc.	5	do	6
143	Everlasting**	latifolius	hp	pink	5	June, Sept.	6
144	white	albus	"	white	5	do	12
145	Persicaria, red	Polygonum orientale	ha	red	6	do	6
146	Polyanthus, fine mixed	Primula elatior var.	hp	var.	1	July, Sept	6
147	Pink, fine clove, mixed	Dianthus moschatus	"	var.	1½	June, July	6
148	annual Chinese	annuus	ha	var.	1	July, Aug.	6
149	Poppy, fine mixed	Papaver somniferum	"	var.	2	July, Sept.	6
150	superb fringed	fimbriata	"	wh. & r.	2	do	6
151	Ranunculus	sinensis	"	var.	1	do	6
152	Phlox, perennial, mix'd	Phlox sp.& va. [driacus	hp	var.	3	do	6
153	Prince's Feather	Amaranthus hypocon-	ha	cr.	2½	do	6
154	Petunia, white	Petunia nyctaginiflora	"	white	2	July, Oct.	6
155	purple	phœnicea	"	violet	2	do	6
156	fine mixed	var.	"	var.	2	do	12
157	Rocket, sweet	Hesperis matronalis	hp	purp.	2	July, Sept.	6
158	Rose Campion	Agrostemma githago	"	purp.	2	June, Sept.	6
159	Scabiosa, fine mixed	Scabiosa, var.	ha	var.	2	July, Sept.	6
160	purple	atropurpurea	"	purp.	2	do	6
161	Sensitive Plant	Mimosa sensitiva	ta	pink	1	do	6
162	Snapdragon, fine mix'd	Antirrhinum majus	hp*	var.	1½	do	6
163	Stock, 10 w. fine mix'd	Mathiola annua var.	hha	var.	1½	June, Sept.	6
164	scarlet	coccinea	"	sc.	1½	do	6
165	white	alba	"	white	1½	do	6
166	purple	purpurea	"	purp.	1½	do	6
167	giant white	alba major	"	white	1½	do	6
168	wall leaved white	var.	"	white	1½	do	6
169	Stock, Ger. 10 w. mix'd	annua densi-	"	var.	1	do	6
170	dwarf Carmine	miniata [flora	"	var.	1	do	6
171	Mulberry	var.	"	mul.	1	do	6
172	new yellow	lutea	"	yel.	1	do	6

JOSEPH BRECK AND CO.'S CATALOGUE OF FLOWER SEEDS.

Number.	Common Name.	Scientific Name.	Duration of Plant	Color of Flower.	Height in feet.	Period of Flowering.	Price.
173	Stock crimson	Mathiola kermesina	hha	cr.	1	June, Sept.	6
174	dwarf rose	rosea	"	rose	1	do	6
175	peach blossom	persicæfolia	"	peach	1	do	6
176	violet	violacea	"	vio.	1	do	6
177	red	rubra	"	red	1	do	6
178	white	alba	"	white	1	do	6
179	chamois	var.	"	buff	1	do	6
180	dark	var.	"	dark	1	do	6
181	cinnamon	cinnamomea	"	cin.	1	do	6
182	Stock, Queen, scarlet	incana	hhb	sc.	2	May, Sept.	6
183	purple	purpurea	"	purp.	2	do	6
184	white	alba	"	white	2	do	6
185	Stock, Brompton, scar.	simplicifoli-	"	sc.	2	do	6
186	purple	purpurea [um	"	purp.	2	do	6
187	white	alba	"	white	2	do	6
188	Victoria	new var.	"	cr.	2	do	12
189	imperial	imperialis	hhp	red	2	do	12
190	Sweet Alyssum	Alyssum maritimum	ha	white	1	June, Sept.	6
191	Sweet Sultan, mixed	Centurea var.	hha	var.	1½	July, Sept.	6
192	purple	moschata	"	purp.	1½	do	6
193	white	alba	"	white	1½	do	6
194	yellow	suaveolens	"	yel.	1½	do	6
195	new blush	crocodylium	"	flesh	2	do	6
196	Sweet William, mixed	Dianthus barbatus[lum	hp	var.	1		6
197	Venus's Looking Glass	Prismatocarpus specu-	ha	blue	1	June, Sept.	6
198	Virginian stock	Malcomia maritima	"	red	½	July, Sept.	6
199	Wallflower, blood	Cherianthus Cheiri	hhp	br.	1½	May, Aug.	6
200	yellow	flava	"	yellow	1½	do	6
201	double dark	flora pleno	"	dark	2	do	12
202	purple	purpurea	"	purp.	2	do	6

The following Seeds, with very few exceptions, have no popular or English name, and in consequence, the scientific names are arranged in alphabetical order. The names in the second column, by which they are often called, are mostly literal translations of the specific name; Example—Clarkia elegans, "elegant" Clarkia, &c.

Number.	Scientific Name.	Common Name.	Duration of Plant	Color of Flower.	Height in feet.	Period of Flowering.	Price.
203	Ageratum mexicanum	Mexican	ha	blue	1½	July, Aug.	6
204	Alonsoa grandiflora	great-flowered	gp*	sc.	1½	June, Sept.	12
205	Amarantus tricolor	three-colored	ta	3 col.	1	July, Sept.	6
206	Argemone mexicana	Mexican	ha	white	3	do	6
207	orchroleuca	yellow	"	yel.	3	do	6
208	Athanasia annua	annual	"	yel.	1	do	6
209	Bartonia aurea	golden	"	yel.	1½	do	6
210	Brachycome iberidifolia	Swan River Daisy	hha	blue	1	do	25
211	Browallia elata [lis	blue	ta	blue	1	do	6
212	Campanula pyramida-	pyr'midal bell-flower	hhp	blue	4	June, Sept.	12
213	alba	white	"	white	4	do	6
214	sp. & var.	mixt peren. sort	hp	var.	2	do	6
215	Loreii	Lore's	ha	blue	1	do	12
216	Centaurea americana	American	"	red	3	Aug. Sept.	6
217	Cladanthus arabicus	Arabian	"	yel.	1	July, Aug.	6
218	Clarkia elegans	elegant	"	lil.	1½	July, Sept.	6
219	rosea	new rose	"	rose	1	do	6
220	pulchella	pretty	"	lil.	1	do	6

Number.	Scientific Name.	Common Name.	Duration of Plant.	Color of Flower.	Height in feet.	Period of Flowering.	Price.
221	Clarkia alba	white	ha	white	1	July, Sept.	6
222	Calampelis scraba**	climbing	hhp*	or.	10	do	6
223	Cleome grandiflora	great-flowered	ha	rose	4	do	6
224	Clintonia elegans	elegant	"	bl. & wh	½	do	6
225	" punchella	pretty	"	bl. & wh.	½	do	12
226	Cobæa scandens**	Mexican	gp*	purp.	10	Aug. Sept.	12
227	Collinsis bicolor	two-colored	ha	wh. & p.	2	July, Aug.	6
228	" heterophylla	various-leaved	"	wh. & p.	2	do	6
229	Calandrinia grandiflora	large-flowered	"	p. & red.	1½	July, Sept.	6
230	" discolor	two-colored	"	rose	1½	do	6
231	Cassia Marylandica	Maryland	hp	yel.	4	June, Sept.	6
232	Commelina cœlestis	sky-blue	ta	blue	1½	July, Aug.	6
233	Cuphea silenoides	Silene-like	ha	purp.	2	do	12
234	Dahlia repens	creeping	hha	var.	1	do	12
235	Didiscus cœruleus	blue	"	blue	2	do	12
236	Digitalis aurea	golden	hp*	or.	2	do	6
237	" lutea	yellow	"	yel.	2	do	6
238	" bicornata	two-horned	"	white	3	do	6
239	Dracocephalum altaiense	Altaic	"	blue	½	June, Aug.	6
240	Elichrysum macranthum	large flowered	hha	pink	3	July, Aug.	6
241	Erysimum Peroffskya-	Peroffsky's	ha	yel.	2	July, Sept.	12
242	Eutoca viscida [num	fine blue	"	blue	1½	do	6
243	Gaillardia Richardsonii	Richardson's	hp*	yel.	2	do	6
244	Gilia capitata	bunch-flowered	ha	blue	2	do	6
245	" tricolor	three-colored	"	3 col.	1	do	6
246	Godetia bifrons	two-fronted	"	p. & c.	1	do	6
247	" concinna	neat	"	p. & r.	½	do	6
248	" Lindleyana	Lindley's	"	wh. & r.	1	July, Sept.	6
249	" quadrivulnera	four-spotted	"	spot	1	do	6
250	" Romansovii	Romanzow's	"	pur.	1	do	6
251	" rosea	rose colored	"	rose	1	do	6
252	" rosea alba	rose and white	"	r. & w.	1	do	6
253	" rubicunda	blushing	"	purp.	1	do	6
254	" viminea	twiggy	"	lil.	1	do	6
255	" vinosa	wine colored	"	purp.	1	do	6
256	" Willdenovil	Willdenow's	"	flesh	1	do	6
257	Helenium Douglasii	Douglass's	"	yel.	1	July, Aug.	6
258	Heliophylla araboides	Arabis-like	"	blue	1	do	12
259	Impatiens candida	white	ta	white	3	July, Sept	12
260	Ipomæa coccinea**	scar. Morning-glory	ha	sc.	10	July, Oct.	6
261	" rubro cærulea**	sky blue	gp	blue	10	do	25
262	" Nil**	blue	ha	blue	10	do	6
263	Lasthena californica	California	"	yel.	1	July, Aug.	6
264	Leptosiphon and rosaceous	long-tubed	"	lil.	1	do	6
265	" densiflorus	dense-flowered	"	var.	1	do	6
266	Limanthes Douglasii	Douglass's	"	yel.	1	do	6
267	Lisianthus Russellianus	Duke of Bedford's	hhp	purp.	1½	June, Sept.	25
268	Loasa Pentlandica	Mr. Pentland's	ha	yel.	6	do	12
269	Lobelia gracilis	graceful	"	blue	¼	July, Oct.	6
270	" cardinalis	Cardinal flower	hb	sc.	½	June, Aug.	6
271	Lophospermum scandens**	climbing	ha	rose	10	July, Oct.	12
272	" atrosanguineum	dark red	"	cr.	10	do	12
273	Lotus jacobæus	dark flowered	ta	black	1½	do	12
274	Lupinus Cruikshankii	Cruikshank's	ha	var.	2	do	12
275	" nanus	dwarf,	"	p. & bl.	1	do	6
276	" Hartwegii	Hartweg's	"	b. & w.	1	do	12
277	" polyphyllus	many-leaved	hp	blue	3	June, Aug.	6
278	" alba	white	"	white	3	do	6
279	Lychnis chalcedonica	scarlet	"	sc.	2	do	6
280	" cœli rosa	Rose of Heaven	ha	flesh	1	do	6
281	Madaria elegans	elegant	"	yel.	2	July, Sept.	6

Number	Scientific Name.	Common Name.	Duration of Plant.	Color of Flower.	Height in feet.	Period of Flowering.	Price.
282	Malope grandiflora	large-flowered	ha	cr.	2	July, Sept.	6
283	alba	white	"	white	2	do	6
284	Malva elegans	elegant	"	sc.	3	do	6
285	zebrina	striped	"	str.	2	do	6
286	Martynia fragrans	fragrant	ta	var.	2	do	12
287	Maurandia Barclayana	Barclay's	hha	purp.	10	July, Oct.	12
288	semperflorens**	ever-flowering	"	rose	8	do	12
289	Mimulus, sp. & var.	fine mixed	"	var.	1	June, Aug.	12
290	Nemophila atomoria	white spotted	"	w. spot	1	July, Oct.	6
291	discoidalis	white bordered	"	black	1	do	25
292	insignis	beautiful blue	"	blue	1	do	6
293	Nolana atriplicifolia	large blue	"	blue	1	do	6
294	Œnothera macrocarpa	long-fruited	hp*	yel.	1	May, Aug.	6
295	tetrapetra	white	ha	white	1	July, Sept.	6
296	Oxyua chrysanthemoi-	Chrysanthemum-like	"	yel.	1	do	6
297	Papaver orientale [des	Oriental poppy	hp	sc.	2	May, June	6
298	Pentstemon, mixed	fine mixed	"	var.	2	June, Aug.	6
399	Phacelia tanacetifolia	Tansy-leaved	ha	blue	1½	July, Sept.	6
300	Phlox Drummondii	Drummond's annual	hha	var	1	July, Oct.	12
301	Portulaca splendens	splendid	"	cr.	½	do	12
302	Thellusonii	Thelluson's	"	sc.	½	July, Oct.	22
303	Potentilla formosa	handsome	hp	rose	1½	June, Aug.	6
304	astrosanguinea	dark red	"	cr.	1½	do	6
305	Rodanthe Manglesii	Mr. Mangle's	hha	rose	1	July, Sept.	12
306	Rudbeckia amplexi-caulis	stem-clasping	ha	yel.	1½	do	6
307	lasciniata	yellow	hp	yel.	2	June, Sept.	6
308	Salpiglossis, fine mixed	fine-mixed	hha	var.	1	July, Sept.	12
309	Salvia coccinea	scarlet	hhp*	sc.	2	June, Oct.	12
310	splendens	scarlet	"	sc.	3	do	12
311	patens	fine blue	"	blue	3	do	25
312	Schizanthus pinnatus	pinnate leaved	ha	l. p. y.	1½	July, Sept.	6
313	humilis	dwarf	"	l. r. y.	1	do	6
314	porrigens	spreading	"	l. p y.	1½	do	6
315	Priestii	Priest's	"	white	1½	do	6
316	pulchella	pretty	"	wh. & pur	1	do	12
317	venustus	beautiful	"	dark	1	do	6
318	retusus	blunt-leaved	"	cr. & yel.	1	do	12
319	Schizopetalon Walkerii	Walker's	hha	white	1	do	12
320	Silene regia	scarlet	hp*	sc.	1½	June, Sept.	12
321	Sophora japonica	Japan	hp	purp.	2	do	6
322	Stevia serrata	sweet-scented	ha	white	1½	July, Sept.	6
323	Thunbergia alata**	winged	ta	buff	6	July, Oct.	12
324	alba**	white	"	white	6	do	25
325	aurantiaca**	orange	"	or.	6	do	25
326	Barkerii**	Mr. Barker's	"	white	6	do	25
327	Tropæolum peregrinum**	Canary-bird flower	"	yel.	10	July, Sept.	12
328	Verbena aubletia	Mr. Aublet's	ha	lil.	1	July, Oct.	6
329	venosa	veined	hhp*	bl.	1	do	6
330	var. & sp.	extra fine mixed	"	var.	1	do	25
331	Zinnia elegans	elegant	hha	vio.	2	July, Sept.	6
332	alba	white	"	white	2	do	6
333	coccinea	scarlet	"	sc.	2	do	6
334	var.	fine mixed	"	var.	2	do	6

General Directions for the Management of Annual, Herbaceous and Climbing Plants.

Annual Flower Seeds should be sown during the month of May, on borders of *light, rich earth, very finely pulverized:* the borders having been previously well dug, arrange with a trowel small patches therein, about six inches in width, at moderate distances, breaking the earth well, and making the surface even: draw a little earth off the top to one side, then sow the seed therein, each sort in separate patches, and cover with the earth that was drawn off, observing to cover the small seeds less than a quarter of an inch deep, the largest in proportion to their size : but the sweet pea and bean kinds must be covered one inch deep. When the plants have been up some time, the larger growing kinds should, where they stand too thick, be regularly thinned, observing to allow every kind according to its growth, proper room to grow.

As a *general principle, almost everything that grows, thrives best in a rich soil;* there are a few exceptions, but they are so trifling, that this rule may be laid down for all practical purposes; therefore make your ground rich; decayed vegetable matter from the woods is best for a flower garden : dig and turn it well over, and make it level; then rake it smooth; if it is well dug, it will be perfectly level, therefore the raking is necessary to make it smooth and fine.

As the stalks of flowering plants shoot up, they generally require thinning, and props for support; and the blossom of a plant or shrub, no sooner expands than it begins to wither, and must be cut off, unless, as in some of the ornamental shrubs, they are left for the sake of the beauty of their fruit.

Always water your plants in the evening—the water then has time to sink into the earth and be imbibed by the plants during the night.

Summer Flowering Bulbs, Dahlias and Plants, for Planting in May.

Gladiolus natalensis, yellow and crimson, showy, each	$ 20
—— cardinalis, scarlet, splendid, each	37½
—— floribundus, pink with dark lines, beautiful, each	37½
—— ramosus, blush, new and very fine, each	1 00
Amaryllis formosissima, velvety crimson, superb, each	25
Tigridia pavonia, light red, spotted, each	12½
—— conchiflora, yellow spotted, each	25
Tuberoses, fine double white, very fragrant, each	25
Commelina tuberosa (or cœlestis,) fine blue, each	12½
Pæonies, of all the finest double varieties, each	50 to 1 00
Dahlias, upwards of 200 varieties, including the finest new ones to be obtained in England, a Catalogue of which is annually published :—per doz.	$2 00 to 8 00
Carnations, extra fine and good varieties, each	25 to 50
Picotees, extra fine, with white and yellow grounds, each	25 to 50
Pinks, fine double sorts, clove scented, &c., per dozen	2 00
Verbenas, (in 20 varieties) for turning out into the border, per dozen	2 00
Chinese Roses, of dwarf kinds for planting in masses, per dozen	3 00
Hardy Herbaceous Plants, in twelve different handsome sorts, per dozen	3 00

Garden Tools and Implements of every description.--Such as Spades, Shovels, Forks, Hoes, Dutch Scuffles, Verge Cutters, Rakes, Garden Reels and Lines, Trowels, Grass Shears, Pruning Shears, Pruning Saws, Pruning Knives, Budding Knives, Grape Scissors, Syringes, &c. &c. Russia Mats of the best quality.

JOSEPH BRECK & CO.
SEED STORE AND AGRICULTURAL WAREHOUSE,
51 & 52 North Market Street,
BOSTON,

OFFER FOR SALE ONE OF THE MOST EXTENSIVE COLLECTIONS OF

VEGETABLE, AGRICULTURAL, AND GRASS
SEEDS

TO BE FOUND IN THE COUNTRY, COMPRISING A LARGE ASSORTMENT, AND INCLUDING

THE NEWEST AND MOST APPROVED VARIETIES,

SUCH AS

Prince Albert and Cedo Nulli Peas, Vanack and other Early Cabbages, Seymour's and Bailey's Superb Celeries, New Brocolis, Prize Cucumbers, &c., &c.

A LARGE AND SUPERIOR COLLECTION OF

BULBOUS FLOWER ROOTS,

AND THE

NEWEST DAHLIAS.

——ALSO——

GARDENING, BOTANICAL AND AGRICULTURAL BOOKS,

—— From their Nurseries ——

AN UNRIVALLED COLLECTION OF

FRUIT TREES,

ORNAMENTAL TREES, SHRUBS AND EVERGREENS,

GRAPE VINES, GOOSEBERRIES, STRAWBERRIES, &c.

GREEN-HOUSE AND HARDY HERBACEOUS PLANTS,

RHUBARB, ASPARAGUS ROOTS, &C.

☞ *Orders filled at one day's notice, and the plants packed and forwarded to all parts of the country.*

NEW ENGLAND FARMER.

This is a weekly paper, devoted to Agriculture, Gardening and Rural Economy, published by JOSEPH BRECK & CO., edited by JOSEPH BRECK.

It is printed in quarto form, paged, making a volume of 416 pages annually, to which a title page and index are furnished gratis. This paper has been published 23 years, and is the oldest agricultural paper, except the American Farmer, in the country; during this time the most assiduous exertions have been made by the editors and publishers to make it acceptable to the Farmer and Horticulturist.

The New England Farmer will contain a weekly report of the sales of cattle at Brighton: the state of the market, crops, &c.

The New England Farmer is published every Wednesday evening, at the low price of $2,50 per annum. A discount of 50 cents will be made when paid in advance.

JOESPH BRECK & CO.

Appendix II

A Chronological Documentation of Annuals through the Trade

EXPLANATORY NOTES:

The annuals are listed alphabetically by genus. To locate them by their common names, see Index.

Descriptions from *Hortus Third* are rearranged and condensed, with emphasis on flower color and form and the plant's general habit. Because current epithets often differ from those found within the chronological limits of this study, this information should also be useful in bringing the reader up to date.

Time periods discussed vary with each annual. They are based on the time lapses between significant developments in the trade with respect to the particular plant in question.

Common and Latin names are given as they appeared in the original lists, including misspellings and inaccuracies.

Descriptions appear in the language of the catalogues. Technical inaccuracies—such as the term "variety" for "cultivar," reference to composites with "double" flowers or "petals" rather than florets, etc.—are included.

Descriptions not in quotation marks are condensed from original versions found in several catalogues; superlatives are generally omitted for brevity.

Quoted material is included to indicate specific views and statements made by particular seed firms or individuals. Abbreviated citations follow in parentheses. Catalogue page numbers, here as throughout the study, are omitted when references appear in alphabetical order.

Flower colors are capitalized only when they were cultivar names (i.e., Snow White, Peach Blossom, etc.).

Initials following each entry are abbreviations for the names of seed firms. They should not be considered as referring to the only catalogues offering the form listed, but rather to the author's initial sources for discovery of the cultivar. A key to these sources follows:

J.B. .... Joseph Breck & Company
T.B. .... Thomas Bridgeman
W.A.B. .... W. Atlee Burpee Company
J.H. .... Joseph Harris Company
P.H. .... Peter Henderson & Company
C.M.H. .... Charles M. Hovey
D.L. .... (David) Landreth's Seed Company
J.T. .... J. M. Thorburn & Company
J.V. .... James Vick & Sons
T.W.W. .... T. W. Wood & Sons

Sand Verbena

Abronia sp.   Nyctaginaceae

DESCRIPTIONS FROM HORTUS THIRD

Abronia arenaria, now *A. latifolia*—flowers lemon-yellow; prostrate. Coast of California to British Columbia.

A. umbellata—calyx rose or rarely white; prostrate. Commonly cultivated. Coast of California to British Columbia.
    cv. 'Grandiflora'—flowers larger.
    'Rosea'—flowers pale.

A. fragrans—flowers white, night-blooming. British Columbia to northern Mexico.

A. villosa—similar to *A. umbellata* but annual. *Var. Villosa*—calyx purplish-rose, opaquely winged with coarse veins. Nevada and California, south to Arizona and Baja California. Desert.

1865–1868

Abronia crux maltae: Noted as a new introduction by C. M. Hovey, "The flowers . . . are in axillary heads on long peduncles, of a deep purplish rose color, the throat swollen and of a bright emerald green, while the tube is pink or flesh color. It was detected in the Carson Valley, Washoe, and first known in 1860." (C.M.H., February 1868, p. 42) Described by W. Robinson (*The English Flower Garden*, 1883) as "a pretty species with white flowers and sweetly scented." Not common in the trade.

umbellata: "A fine half hardy annual, with clusters of sweet-scented flowers, resembling the Verbena; rosy lilac; 6 inches in height. Fine for baskets and desirable in the garden." (Vick, 1865) (J.B., C.M.H., J.V., P.H.)

1869–1872

A. arenaria: pure waxy-yellow, scented, new (P.H.)

fragrans: white (J.V.)

1873–1880

A. arenaria: offered in catalogues with more extensive lists (P.H., J.B., J.V.)

fragrans: P.H. offered as a novelty in 1880; snow white, vanilla fragrance

umbellata: most commonly offered (J.B., J.H., P.H., D.L., J.V., W.A.B.)

1881–1900

Landreth's began to stop listing Abronias.

A. umbellata: still offered by above firms, as well as J.T.

umbellata grandiflora: large flowered (J.B.)

villosa: purple (J.B.)

1901–1914

Abronias begin to appear in the back pages of catalogues as specialty plants; recommended for borders, rockeries, baskets.

CHINA ASTER

Callistephus chinensis   Compositae

DESCRIPTION FROM HORTUS THIRD

Callistephus chinensis—erect to 2½ ft., stems branching, hispid-hairy; upper leaves spatulate; heads showy, to 5 in. across; disc flowers yellow, ray flowers in 1–2 rows in wild plants. China.

There are many cvs., varying in height, season of bloom, and composition of the head; the ray flowers may be very numerous, and frequently replace most of the disc flowers, showing a wide range in size, shape, and color—violet, purple, blue, red, pink, white, and even pale yellow.

1865–1870

Even at this time, China asters were known by "fancy" names. The following lists are exclusively in this form.

Chrysanthemum-flowered, Tall: fine large flowers (J.B., D.L., J.V.)

Chrysanthemum-flowered, Dwarf (new): growing only about one foot in height, with large, very perfect flowers; a free bloomer; later than other varieties; 12 mixed colors.
—Snowy White: changing from white to azure blue as the flowers become old; every one of the flowers perfect. (J.V.)

Cocardeau, or New Crown: very double, the central petals of a very pure white, sometimes small and quilled, surrounded with large flat petals of a bright color, as crimson, violet, scarlet, etc.; 18 in. Also sold separately in carmine, violet, blue, deep scarlet, violet-brown, and rose (J.B., J.V.)

Dwarf Bouquet, Newest: each plant is covered with bloom, only an occasional leaf being visible, as in a well-arranged bouquet. About a dozen different colors. (J.V.)

Giant Emperor, New: enormous flowers, generally double, truly giant. "At first this did not meet our expectations, but each importation has shown improvement." (Vick, *Floral Guide*, 1865) Lilac, violet, peach, brown, blue, carmine, rose, red, etc., in separate colors.
—Snowy White: this has proved excellent; flowers the purest white, of enormous size, and good form. (J.B., J.V.)

Hedge-Hog, or Needle: petals long, quilled, and sharply pointed; very curious and fine; 2 ft. (J.V.)

Imbrique Pompon(e): almost a perfect globe, imbricated; 18 in. 12 colors. (J.B., P.H., J.V.)

Original Chinese: plant tall, flowers large and loose; distinct in appearance and of bright colors. Released by the Vilmorin Co. (J.B.)

Peony-flowered Globe, New: earliest of the Asters—at least two weeks earlier than Truffaut's Peony-flowered; flowers very large; plant branching and strong, and does not require tying. (J.V.)

Pyramidal Bouquet, Dwarf: about 10 in. in height; abundance of flowers; very early. (J.B., J.V.)

Pyramidal-flowered German: extra late, branching, good habit, fine grower, needs no tying. (J.V.)

Ranunculus-flowered Bouquet, Dwarf: small, very perfect flowers; most profuse bloomer; 1 ft. (J.V.)

Reid's, New: the finest quilled Aster grown. None of the quilled varieties are as fine as those with flat petals. (J.V.)

Rose, Large-flowered, or La Superbe: large, bright rose flower, often more than four in. in diameter, of the peony-flowered class; 20 in. (J.V.)

Truffaut's Peony-flowered Perfection: petals long, a little reflexed; 20 in. to 2 ft.; mixed colors. (J.B., P.H., J.V.)

Victoria, New: carmine-rose, flowers as large as the Emperor Aster, habit pyramidal, nearly 2 ft. high, each plant bearing from 20 to 40 flowers. This was a novelty of 1863. Somewhat of the character of La Superbe. (J.B., J.V.)

1871–1875

Goliath: large flowered; mixed colors. (P.H.) "It is no better than the old Giant Emperor for America, and has the same faults." (Vick, *Floral Guide*, 1875)

Schiller, New: a late, dwarf bouquet Aster, of peculiar habit and great beauty. Height about 15 in., with great quantity of bloom; mixed. (J.B.)

1876–1883

Burpee and Landreth offered all major types listed above for the period 1865–1870.

Betteridge's Choice: quilled; mixed colors. (J.H., J.V.)

Multiflora Mauve, Half-Dwarf: flowers perfect and abundant, delicate white and mauve; about 15 in. French origin. (J.B.)

1884–1887

Boston Florists' White: largely used for cutting by city florists. Snowy white; finely imbricate; specially saved seed. (J.B.)

Comet: "This represents a new and beautiful section, of the same height and habit as the Dwarf Paeony Perfection Asters, forming fine, regular pyramids 12–15 in. high, and covered profusely with large double flowers. The shape differs from that of all classes of Asters in cultivation, and resembles very closely that of a large-flowered Japanese Chrysanthemum. The petals are long and somewhat twisted or wavy and curled, recurved from the centre of the flower outwards in such a regular manner as to form a loose but still dense semi-globe. Well grown plants produce from 25–30 perfectly double flowers measuring 3½–5½ in. in diameter. The colour is a delicate pink bordered with white. The variety was raised by Messrs. Haage & Schmidt, of Erfurt" ("Continental Novelties, Aster Comet," *Gardener's Chronicle*, December 25, 1886). Appeared in Breck's Novelty Supplement, 1887.

Lilliput: a very fine, dwarf Aster; Benary. (J.B., W.A.B.)

Mignon: pure white; like Victoria in habit; Benary. (J.B.)

Washington, New: without exception the largest Aster in cultivation, the flowers being frequently 4–5 in. in diameter, mixed colors; 2 ft. (J.B.)

Zirngiebel's Improved Double White: cross between Truffaut's Perfection and Victoria; 2 ft.; pure white, beautifully imbricated. (J.B.)

1888–1894

Comet: rose and white. (J.T.) carmine, giant white, light blue, blue bordered white, lilac bordered white, pure lilac. (W.A.B.)

Diamond Asters: especially recommended by the floral committee of the R.H.S. flowers 2–2½ in. across, double, petals incurved; particularly valuable for cutting; the plant grows to 20 in., the main stalk stiff, upright growth with numerous stems starting from the base and each again branching and terminating with perfect flowers of rich and varied colors. Offered in separate colors: pure white, deep violet, deep carmine, purplish-lilac, pink and white, dark crimson, rose, crimson with white, deep violet with white, reddish-violet (W.A.B.)

Harlequin: upright, medium height, profuse blooming, a new race of odd, striking flowers.
—Red: half pure white and the balance bright red
—Blue: pure white irregularly interspersed with deep blue petals (W.A.B.)

Imbricated Yellow: lemon or light yellow flowers, "the first approach to yellow in Asters that we have considered worthy of note." (Burpee, *Farm Annual*, 1888)

Queen Asters, Large-flowering Dwarf: flowers of extraordinary size, perfectly double, resembling the Victoria in form. Originated in Germany.
—White: spotless white
—Crimson: delicate rosy-crimson (W.A.B.)

Queen of the Market: graceful spreading habit, flowers produced on very large stems, useful for cutting; early flowering, two weeks earlier than any other Aster; deep blue, pure white, light pink, and deep rich rose (W.A.B., J.T.)

Shakespeare: very dwarf, continues in flower a long time, in mixed colors (W.A.B., J.T.)

1895–1905

The following appeared in addition to the standard China asters listed above (source: J. M. Thorburn & Co.):

Ball or Jewel: long stemmed, round flowers, good for cutting or bunching, dwarf

Branching: broad, handsome bush habit, with large, long stemmed and long petaled flowers; feathery

Empress Frederick: pure white; for pots

Hohenzollern: white and rose

Japanese Tassel: curiously waved, long petals, resembles Japanese chrysanthemums

Lady Aster: pure white; graceful habit, peculiar straight edged leaves

Ray: white

Snowball: white

Triumph: incurved, brilliant deep scarlet

1906–1914

Eugenie: imbricated pompon, mixed colors (T.B.)

Favorite: outer petals wavily reflexed, twisted; color blush on opening, changing to deep rich pink (D.L.)

King Humbert: dwarf, gigantic comet-like flowers, branching (D.L.)

Semple's Branching: late blooming, in separate colors—Shell-Pink, Lavender-Blue, Crimson (T.B.)

Vick's Strains of China asters:
    Daybreak: round flowers, sea-shell pink
    Early Branching: flowering two weeks earlier than late branching sorts, in two colors—white and rose
    Late Branching: "It is as favorably known among European as American growers. . . . It begins blooming about August 15th and continues throughout the season. The flowers are of extraordinary size, and are borne on long, graceful stems from fifteen to twenty inches in length." (Vick, *Garden and Floral Guide*, 1908) —snowy-white, pink, rose, crimson, purple, dark violet, lavender, mixed
    Purity: identical to Daybreak but pure white
    Rosy Carmine, Branching: popular with florists
    Sunrise: oval flowers, 4 in. in diameter, delicate pink; discovered in a field of Daybreak asters
    Sunset: light pink shading to deep pink in the center; globe-shaped flowers born on long stems

COCKSCOMB

Celosia cristata Amaranthaceae

DESCRIPTION FROM HORTUS THIRD

Celosia cristata—a tetraploid cultigen similar to *C. argentea*, enlarged spikes variously crested, plumed, or feathered. Flowers from white to yellow, purple, and shades of red. Many cvs., most fall into the Childsii Group with ball-like knobs at end of each branch. Nana Group—plants dwarf. Plumosa Group—feathered spikes. Spicata Group—flowers silvery-rose in slender spikes.

1865–1872

Celosia cristata: crested cockscomb. Offered in separate colors—Crimson Dwarf, Yellow Dwarf, Violet Dwarf, Scarlet Giant, Tall Violet, Tall Rose, Tall Sulphur, Dwarf and Tall mixed (J.V.)

nana aurantiaca pyramidalis: bright, fawn-colored panicles, fine foliage (J.V.)

pyramidalis argentea: pyramids of feathery spikes, silver white, shaded with bright rose; will keep like Everlasting flowers if cut young; 3 feet high. (J.V.)

—*aurantiaca*: scarlet tipped with orange; 3 feet. (J.V.)

—*coccinea*: spikes very large, feathery, scarlet, exceedingly showy; 3 feet (J.V.)

pyramidalis versicolor: light crimson flowers, verging on crimson violet

—*foliis atrobruneis*: foliage reddish-brown, panicles golden-orange
spicata rosea: spikes of rose-colored flowers that keep well for winter ornaments (J.V.)

1873–1879

C. cristata aurantiaca: orange panicles (D.L.)

cristata variegata: "showing a mixture of red and yellow, and hardly worth culture; very late, and does best South." (Vick, *Floral Guide*, 1875)

japonica, or New Japan Cockscomb: "This is an entirely new variety of Cockscomb, received from Japan last year. It is far better and more brilliant than the old variety a single plant being an object of great beauty, while a bed containing a dozen plants is not equalled for a garden display by anything we are acquainted with. The branches, from the roots to the small leaf-vein, are scarlet or crimson, the combs are almost as delicately cut as ruffled lace, often in pyramidal masses, while the colors are of the brightest description imaginable." (Vick, *Illustrated Floral Guide*, 1873)

superba plumosa: feathered, bright crimson (J.B., J.V.)

1880–1888

C. cristata nana: dwarf, scarlet (J.B., D.L.)

> *cristata pyramidalis*: pyramidal habit, mixed colors (J.B., D.L.)

Giant Empress: mammoth, bright purple combs and dark bronze foliage (W.A.B.)

Glasgow Prize or Tom Thumb: dwarf, with dark leaves and crimson combs (W.A.B.)

Variegated (see *C. cristata variegata* under 1873–1879 above): bright yellow and crimson (W.A.B.)

1889–1893

C. cristata candidissima compacta: silvery foliage, dwarf (D.L.)

> *Huttoni*: dark foliage, crested (J.T.)

Moss Head: crested, mixed colors (J.T.)

President Thiers: bright red, dwarf (J.T.)

1894–1907

New Plumed Celosia: Triumph of 1889 Paris Exhibition. Bronze foliage, fiery flowers (W.A.B.)

Queen of Dwarfs: dark scarlet, crested; 8 inches high (W.A.B.)

1908–1914

C. cristata Dwarf Chamois: crested sorts in separate colors—Fawn, Copper/Bronze, Dark Crimson, Empress (purple), Glasgow Prize (crimson), Golden Yellow, Queen (rose), Scarlet, Violet, Vesuvius (scarlet), mixed (J.T.)

> *plumosa*: feathered sorts in separate colors—Crimson, Golden Yellow, Thompson's Superb (crimson), Ostrich Feather (crimson), Orange, mixed (J.T.)
>
> *spicata* (*argentea linearis*): green and white (J.T.)

CLARKIA

Clarkia sp.   Onagraceae

DESCRIPTIONS FROM HORTUS THIRD

Clarkia elegans, now *C. unguiculata*—flowers lavender-pink to salmon or purple with long slender basal claws, petals triangular to rhombic. California. Cvs. with double flowers and many colors.

C. pulchella—Petals bright pink to lavender, three-lobed, the lateral lobes usually narrower than the middle one. Rocky Mountains to Pacific Coast.

C. grandiflora, now *C. coccinea*—deep bright pink.

1861

Double Clarkia at London Shows: "Messrs. Carter & Co., of Holborn, exhibited some specimens of a new Clarkia, very distinct and beautiful, much brighter in color than any of the older varieties; a rich rosy pink, and apparently quite constant in its double properties. A figure of this will appear in the *Floral Magazine*. For this a 1st Class Certificate was awarded." (Meehan's *Monthly*, vol. 9, September 1861, p. 283)

New Double White Clarkia elegans: "Vilmorins now offer a double white. There has been a double rose before. Only a portion of the stamens are transformed into petals, so that enough pollen is produced to fertilize the stigma and enable the double variety to reproduce itself from seed." (Meehan's *Monthly*, vol. 1, January 1861)

1865–1868

Clarkia: "A showy and interesting class of hardy annuals that flower freely, with a good variety of delicate colors, and form a cheerful and attractive bed. They do not bear our hot summer suns very well, and it is best to give them the benefit of a little shade. They will then flower magnificently during the autumn months, even after pretty hard frosts. About one foot high." (Vick, *Illustrated Catalogue and Floral Guide*, Spring 1865, p. 16)

pulchella: pretty, large-flowered; mixed colors
—*pulcherrima*: rose-violet
—*marginata*: rose-crimson, edged with pure white (also called
   —*magenta*, J.B.)
—*integripetala*: large flowers, crimson, new
   —*marginata*: rosy purple, edged white, new
—*alba*: white, desirable for contrast

—*flore-pleno*: double, very beautiful; rich magenta 18 inches high, new (also, *rosea plena*, J.B.)

elegans alba flore-pleno: double white, new
—*flore-pleno carnea*: double, flesh-colored
—*rosea*: double rose
—*violacea*: double violet

(The above list, offered by James Vick and Joseph Breck, was the most extensive found. Most catalogues of this period offered *C. elegans* mixed in single and double forms.)

In addition to the above list, Breck also offered:
pulchella Tom Thumb: rosy purple, dwarf
—*compacta*: very dwarf and compact

1869–1875

Peter Henderson offered most of the forms listed above.

Vick's list was reduced to *C. pulchella* (mix), —*integripetala*, —*flore-pleno*; *C. elegans alba flore-pleno*,—*violacea*. His list was further reduced in 1873 to Clarkia, Double mix; Single mix.

1876–1883

C. elegans: mixed (W.A.B., J.H.)

Clarkia, Mixed: "They are more neat and cheerful than showy and generally please, but suffer under our extreme heat, recovering as the autumn approaches. They do best when sown in August or September and sheltered during winter." (*Landreth's Illustrated Catalogue of Flower Seeds*, 1879)

1884–1890

Clarkia Mrs. Langtry: "Petals have an even edge. Pure white . . . evenly defined disk or center of brilliant carmine crimson. Dwarf. Flowers possess unusual substance. Pot or bed." (Breck's *Flower Seed*, 1884). Mrs. Langtry was also offered by Burpee and Henderson.

Clarkia Tom Thumb: white, dwarf (J.T.)

1891–1900

C. elegans nana rosea: dwarf, rose (Breck & Co. trial grounds)

Clarkia Purple King: purple, large-flowered (J.B.)
—Salmon Queen: salmon-pink, large-flowered (J.B.)

1901–1914

C. pulcherrima alba nana: white, ½ foot (T.B.)

COLLINSIA OR CHINESE HOUSES

Collinsia sp.   Scrophulariaceae

DESCRIPTION FROM HORTUS THIRD

Collinsia bicolor, now *C. heterophylla*—flowers two lipped; lower lip violet or rose-purple, upper lip white. Northern California to northern Baja California

cvs. 'Alba' and 'Candidissima' have white flowers.

1865–1871

Collinsia: "a delicate, pretty, free-blooming genus . . . not very showy." (James Vick, Spring 1865)

C. multicolor marmorata: white and rose marbled, new, one foot in height (J.B., J.V.)

bicolor: purple and white (J.B., J.V.)

grandiflora: blue, white, and lilac (J.B., 1868)

multicolor bartsiaefolia: purple lilac (J.B., 1868)

1872–1878

Peter Henderson offered *C. bicolor*; *C. grandiflora*.

C. violacea: "more compact and bushy, 9–12 inches high. flowers smaller than *C. verna*, better form. Upper lip nearly pure white, lower lip of a beautiful deep violet blue. Does well even when planted in early spring. Awarded several 1st Class Certificates last spring. (Joseph Breck's *Flower Seed*, 1872)

1879–1889

Collinsia Mixed: "well adapted to clumps or masses near the eye, as it only rises about one foot, flowers of various hues, grows readily from seed; quite a hardy annual." (David Landreth, *Catalogue*, 1879).

Mixed (W.A.B.)

1890–1905

Breck reduced his list to Collinsia Mixed.

C. bicolor, *C. bartsiaefolia*, *C. alba*, *C. candidissima*, *C. grandiflora*, *C. multicolor*, —*marmorata*, *C. verna*, Mix (J.T.)

1906–1914

C. bicolor alba: white flowers (T.B.)

vera: sky-blue and white (T.B.)

CALIFORNIA POPPY

Eschscholzia californica   Papaveraceae

DESCRIPTION FROM HORTUS THIRD

> *Eschscholzia californica*—stems branched, more or less glaucous, usually glabrous, to 2 ft; flowers from deep orange to pale yellow, petals ¾–2½ in. long. Widely distributed in California, where it is the state flower. Variable in growth form and in color of flowers; some forms have received cv. names.
> cvs.: 'Alba'—florets cream-white. 'Aurantiaca'—florets orange. 'Compacta'—of compact habit. 'Crocea'—florets deep orange. 'Rosea'—florets salmon-pink.

E. maritima, now *E. californica*.

1865–1870

Eschscholzia alba or *E. crocea alba*: creamy white (J.B., J.V.)

> *auriantia* (see 'Aurantiaca' above): very deep orange which does not fade as the *crocea*. German introduction (J.B.)
>
> *californica*: bright yellow, darker in center (J.V.)
>
> *compacta*: yellow and orange (J.B.)
>
> *crocea*: rich orange, darker in center (J.B., J.V.)
>
> *rosea* or *alba rosea*: tinged rosy pink, may eventually become a red (J.B.)
>
> *tenuifolia*: flowers small, pale yellow, resembling the primrose, and numerous; a miniature plant only 6 in. high (J.B., J.V.)

1871–1879

E. crocea striata: flowers orange, striped with lemon (J.V.)

> *dentata aurantiaca*: each petal has its edge lapped upon itself, with a mark of deeper color running up the center, orange; edges of petals toothed (J.V.)
>
> *dentata sulphurea*: petals sulphur (J.V.)

1880–1889

E. californica alba flore pleno: white, from original double yellow. Originated in England in 1881. (P.H.)

> *crocea flore pleno*: bright orange-scarlet, shading off to salmon-red (P.H., J.T.)
>
> *crocea striata* (D.L., W.A.B.)

Mandarin: inner side of petals rich orange, outer side bright scarlet (J.B., P.H., J.T., J.V., W.A.B.)

Rose Cardinal: rich rose-purple (W.A.B., J.B.)

1890–1900

Cross of Malta, *E. maritima*: bright yellow, dark orange cross; also Maltese Cross (J.B., P.H., W.A.B.)

E. Douglasii: early flowering habit, pure yellow with golden center, glaucous blue foliage (P.H.)

Hunnemania or Giant: yellow (J.B.)

1901–1914

Burbank's New Crimson California Poppy: very showy, crimson, rare. Also issued as Carmine King (Carter and Co., England). Luther Burbank introduction. (D.L.)

Golden West: light canary yellow flowers with orange blotch at base of each petal forming a cross in center. Overlapping petals delicately waved (D.L.) deep yellow or light orange-yellow (J.T., J.V.)

Rose Cardinal (T.W.W.)

GAILLARDIA OR BLANKET FLOWER

Gaillardia sp. Compositae

DESCRIPTION FROM HORTUS THIRD

Gaillardia picta, now *G. pulchella* var. *picta*—Disc flowers yellow or with red tips. Rays red, tipped yellow or entirely yellow or red. Coastal Virginia to Florida, west to New Mexico and Mexico, north to Colorado, Nebraska, Missouri; Texas.
cvs. 'Indian Chief;' 'Red giant;' and 'Lorenziana' (all disc florets).

1865–1874

"A very valuable class of plants. The prevailing colors are brownish-red, yellow, and orange. They are constant bloomers giving a magnificent display. Few flowers in the garden will attract more attention if planted in beds or masses. They flower early and continue until frost. Half-hardy annuals. Will bear transplanting well." (Vick, *Illustrated Floral Guide*, Spring 1865, p. 21)

Gaillardia alba-marginata: red, bordered with white (J.V., P.H.)

aristata: yellow (J.B.)

coccinea nana: crimson, dwarf (P.H.)

hybrida grandiflora or *grandiflora hybrida*: rich crimson and yellow (J.V., J.B., P.H.)

Josephus: very brilliant, red and orange (J.V.)

picta, or Painted: brownish-red, bordered with yellow (J.V., J.B.)

Richardsonii: orange and brown center (J.B.)

1875–1883

"The Gaillardias are natives of Texas and other Southern States, and are known by the common name of Blanket Flower in some sections of the South, under which name we have received many specimens of seeds and flowers. . . . The double-flowering variety is particularly interesting." (Vick, *Floral Guide*, 1875)

G. *hybrida grandiflora*: purple center, crimson rays, deeply margined with yellow (D.L.)

picta (W.A.B.)

1884–1891

G. *pulchella Lorenziana*: Novelty. "Bloom is brilliant beyond description. Almost ball shaped-flowers are wonderful perfection, sometimes reaching enormous size of 3 inches across. 10 brilliant colors." (Burpee, 1884)

Double flowering, sulphur and golden, yellow, orange, claret, amaranth (P.H., D.L.)

"It is an introduction of great importance, and cannot fail to find a place in every garden." (Burpee, *Farm Annual*, 1888, p. 102)

1892–1914

G. *grandiflora folius aurea*: variegated (D.L.)

> *pulchella Lorenziana* continued in importance. Available commonly. Offered also as *G. picta Lorenziana* and Double Gaillardia.

GILIA OR BIRD'S EYES

Gilia sp. Polemoniaceae

DESCRIPTIONS FROM HORTUS THIRD

Gilia achilleifolia—flowers in dense, terminal, fan-shaped heads, corolla violet or blue-violet; erect annual, to 3 ft.; Southern California, Baja California.

G. capitata—flowers blue, violet, or white, 50–100 in terminal heads to about 1 in. across. British Columbia to central California, east to western Idaho.

G. laciniata—flowers rose, pale lilac to white, in few-flowered clusters. Peru, Chili, Argentina.
var. *erecta*—leaves somewhat fleshy.

G. tricolor—erect, to 2½ ft., flowers in few-flowered, terminal cyme, corolla with lilac or violet lobes, yellow tube, and throat marked with purple. California.

1865–1870

Gilia: "Very pretty when grown in masses, but not showy." (Vick, 1865)

Gilia achillaefolia: purple lilac (J.V.)
—*alba*: white (J.V., J.T.)

capitata: sky blue (J.V.), celestial blue (J.B.), azure blue (J.T.)
—*flore-albo*: white (J.V.)

laciniata: fine blue (J.B.)

liniflora (now *Linanthus liniflorus*): branching, deeply palmate foliage, large white flowers (J.B.)

nivalis: dwarf, white flowers (J.V.)

tricolor: rose, yellow, purple (J.V., P.H.); white, blue, purple (J.B.)
—*flore-albo*: white (J.V.)
—*flore-rosea*: rose (J.V.)

1871–1880

G. rosea splendens: rose (P.H.); Henderson also listed those listed above.

Gilia: "Neat, unpretending, and sure to please; long in bloom; of easy culture; and best seen when reared in clumps, place near the eye. Best in Philadelphia: *G. capitata* and *G. tricolor*; not good in Philadelphia: *G. nivalis* and *G. rosea*." (Landreth's *Illustrated Catalogue*, 1879)

1881–1890

G. achillaefolia, *G. nivalis*, *G. tricolor*: "pretty dwarf annuals. Pots or rockery. Do not transplant well." (Breck, 1886)

capitata, *G. alba*, *G. laciniata*, *G. linifolia*, *G. nivalis*, *G. tricolor*, *G. rosea splendens*: "for beds or rockeries." (James Thorburn, 1890)

Gilia Mix: "fine for cutting; of dwarf habit; will bloom almost anywhere and are fine for massing." (Burpee's *Farm Annual*, 1883)

1891–1914

Gilia: "fine for rockwork" (Landreth, 1891); "old fashioned favorites" (Breck, 1900)

G. tricolor, *G. nivalis alba*, *G. capitata* (T.B.)

BALSAM

Impatiens balsamina   Balsaminaceae

DESCRIPTION FROM HORTUS THIRD

Impatiens balsamina—flowers axillary, short-pedicelled, overtopped by leafy shoots, 1–2 in. across, white to yellow or dark red, often spotted, spur of variable length, incurved. Many forms, chiefly double-flowered, are offered. India, China, and Malay Peninsula.

1865–1879

Aurora: delicate pink, shaded a bluish tinge (J.B., J.V.)

Camellia-flowered: flowers spotted, or variegated, mixed colors with dashes of white. Landreth considered this a class. (J.B., D.L., J.V.)

Carnation-flowered: flowers double, striped like a carnation. Landreth considered this, too, to be a class. (D.L.)

Double Dwarf Rose-, Camellia-, and Carnation-flowered (J.B., D.L., J.V.)

Isabella: pale rose, changing to yellow (J.B., J.V.)

Rose-flowered: double flowered, self-colored balsams (J.B., D.L., J.V.)

Solferino: Carnation type, striped and spotted (J.B.)

1880–1888

Burpee's Superb Camellia-flowered: new strain of large-flowering (2 in. in dia.) double balsams. Mixed and in separate colors—apple blossom, crimson, flesh color, lilac, peach blossom, scarlet, violet, and also spotted in crimson, purple, and white (W.A.B.)

Camellia-flowered Light Lemon: large flowers, light lemon shade (W.A.B.)

Maiden Blush: pale pink blossoms, from American grower and originator of White Perfection (W.A.B.)

White Perfection: immense snowy-white, double flowers (W.A.B.)

1889–1900

Burpee's Defiance Balsams, New: in mixed shades.

Burpee's Fordhook Fancy and Tricolor: improved strains, mixed colors.

Thorburn offered balsam in separate colors—apple blossom, crimson (self and spotted), light citron, Paris white, peach blossom, purple, rose, scarlet, and solferino.

1901–1914

Defiance: mixed colors (D.L.)

Kermesina: double scarlet (T.B.)

Pink Perfection: pink, self-colored (D.L.)

Queen: bright rose (D.L.)

Scarlet Splendens: deep scarlet (D.L.)

Sunshine: scarlet, double (D.L.)

Victoria: double striped (T.B.)

SWEET PEA

Lathyrus odoratus   Leguminosae

DESCRIPTION FROM HORTUS THIRD

Lathyrus odoratus—climbing to six ft., lightly pubescent; leaflets in 1 pair, to 2 in. long; flowers fragrant, in many colors, to nearly 2 in. across, 1–4 on a peduncle. Italy.
var. *nanellus*, L. H. Bailey. not climbing, the plants very compact.

Sweet peas were known by their "fancy" names and are listed accordingly below.

1865–1870

Blue Hybrid: white with blue edge (J.V.)

Painted Lady: rose and white (J.V., P.H.)

Purple (J.B., P.H., J.V.)

Purple Striped: purple and white (P.H., J.V.)

Scarlet Invincible: deep scarlet (P.H., J.V.)

White (J.B., P.H., J.V.)

Yellow (J.V.)

1871–1879

Black: very dark, brownish-purple (J.V.)

Blue Edged: white and pink, edged with blue (J.V.)

Butterfly: white ground delicately laced with lavender-blue (P.H.)

Violet Queen: dwarf habit (P.H.)

1880–1887

Adonis: bright rosy-carmine, a pleasing contrast to White, Invincible Scarlet and other more decided shades already in cultivation (J.B., W.A.B.)

Black Invincible: dark purple (W.A.B.)

Bronze Prince: upper petals rich bronzy maroon, lower petals deep, bright blue (W.A.B.)

Cardinal: robust; producing a great profusion of bright, shining crimson-scarlet flowers, very distinct and handsome; Eckford introduction. (J.B., W.A.B.)

Crown Prince of Prussia: bright blush (W.A.B.)

Duchess of Edinburgh: the standard light scarlet, flushed with crimson, slightly marbled or splashed at the edge with creamy white; wings deep rose. Eckford. (J.B.)

Grand Blue: deep blue, with pale blue center and wings. Eckford. (J.B.)

Imperial Blue: bright blue wings, slightly shaded with mauve, the standard being rich, purplish crimson. Eckford. (J.B.)

Indigo King: dark maroon-purple standard, clear indigo-blue wings. Eckford. (J.B.)

Invincible Carmine: intense crimson-carmine; large and brilliant flowers. Awarded 1st class certificate by R.H.S. Floral Committee. Laxton. (W.A.B., J.B.)

Isa Eckford: creamy white, heavily suffused with rosy pink. Eckford. (J.B.)

Miss Ethel: pink, with carmine blush wings. Eckford. (J.B.)

Orange Prince: bright orange pink standard, flushed with scarlet; bright rose wings veined with pink. Awarded 1st class certificate by R.H.S. Eckford. (J.B.)

Princess Beatrice: rich carmine rose, shaded with lighter and darker tints. Hurst. (J.B.)

Princess Louise: rosy pink standards; deep lilac-blue wings (W.A.B.)

Princess of Wales: shaded and striped with mauve on a white ground; flowers of great substance and perfect shape. Awarded 1st class certificate by R.H.S. Eckford. (J.B.)

Queen of the Isles: bright scarlet, mottled with white, wings flaked with rosy-purple. Eckford. (J.B.)

The Queen: light rosy pink standard contrasting with light mauve wings. Eckford. (J.B.)

1888–1890

Apple Blossom: bright pinkish rose, wings blush; true apple blossom color. Eckford. (J.B.)

Boreatton: bronzy-crimson standards; crimson-purple wings shaded with rose. Eckford. (J.B.)

Captain Clark: tricolor (J.B., W.A.B., J.T.)

Delight: white wings; standards white crested with crimson; small, dwarf habit (J.B.)

Fairy Queen: milky-white standards; wings tinted pink (W.A.B.)

Queen of England: new white variety of extra large size and good substance. Eckford. (J.B.)

Splendour: rich, bright, pinkish-rose shaded with crimson; large and of the finest form. Eckford. (J.B.)

Vesuvius: velvety violet standards shading off into lilac toward the edge; wings are distinctly spotted on a rosy ground color shading into purple at the throat. (W.A.B., J.T.)

1891–1896

American Belle: distinctly spotted; clear, bright rose standard; crystal white wings with purplish carmine spots; desirable for cutting. (W.A.B.)

Blushing Bride: rose and white; grown by Boston florists. (J.B.)

Countess of Radnor: pale mauve standards with a deeper shading of mauve; pale lilac wings (J.B.)

Cupid: dwarf habit, 6 in. high; pure white. Burpee introduction. (J.B., W.A.B.)

Harvard: deep crimson; vigorous and heat resistant (J.B.)

Miss Blanche Ferry: compact growth, requiring no support; pink and white flowers of immense size (J.B.)

Primrose: standards and wings pale primrose yellow; awarded a 1st class certificate by R.H.S. Eckford. (J.B.)

From the 1890s onward, lists of sweet peas became too unwieldy to treat with any degree of accuracy here.

Petunia

Petunia × *hybrida*   Solanaceae

DESCRIPTION FROM HORTUS THIRD

> *Petunia* × *hybrida*—cultigen; apparently a complex of hybrids involving *P. axillaris*, *P. inflata*, and *P. violacea*: flowers 2–3½ in. long, corolla tube fun nel form. Differing from original species in its much larger and variously formed flowers in many colors, and its more stocky growth.
>
> Various cvs. have flowers varying in size, form, and color, sometimes measuring 4–5 in. across, often deeply fringed or full-double, ranging from white to deep red-purple, variously striped and barred, or with starlike markings radiating from the throat; these often are designated in catalogues by such horticultural names as *P. fimbriata*, *P. grandiflora*, *P. multiflora*, *P. nana*, and *P. superbissima*.
>
> This hybrid complex comprises all the highly developed garden petunias.
>
> Certain cvs. may be maintained only through closely controlled breeding and are best propagated by cuttings, others come reasonably true from seed.

The following "cultivars" and "varieties" should be read as forms of *P.* × *hybrida*.

1865–1870

Buchman's New Hybrid, or Blotched: finest mix saved from named varieties (J.V.)

Countess of Ellesmere: rosy-carmine with white throat (J.B., J.V.)

flore pleno, or Double: double flowers in mixed colors; seed produced sparingly and very expensive (J.V.)

Inimitable: red margined and blotched with white (J.B.)

kermesina grandiflora, or —*splendens*: white with purple or crimson throat, blotched with purple or violet (J.B.)

Large-flowered: dark red, purple with green edge (J.B.)
    —*marginata* (J.B., J.V.)

maxima alba: large white flowers (J.B.)

picturata: dwarf; large flowers, velvety scarlet-crimson marbled with white (J.B.)

rosea grandiflora: large flowers, bright rose with white throat (J.V.)

venosa grandiflora: various colors, veined (J.V.)

1871–1878

Blotched and Striped: in various colors (J.B., P.H., J.V.)

Henderson offered Double petunias for cutting, pots, and as florists flowers.

Vick's Fringed: a new strain, with fringed and frilled edges; coming unusually true from seed (J.V.) 50 seeds for 50¢.

1879–1884

fimbriata flore pleno, or Double Fringed: various colors; imported from European florists (J.H., P.H.)

grandiflora fimbriata: large single flowers, white blotched with crimson (W.A.B., J.B.)

Lilliput: flowers spotted, striped or self-colored; resembling Salpiglossis (J.B., W.A.B.)

nana compacta multiflora, or Dwarf Inimitable: cherry red with white star; compact habit (J.B., W.A.B.)

robusta flore pleno: semi-dwarf habit; double flowers. Benary originator (J.B., P.H.)

1885–1894

Black-throated Superbissima: erect habit; black throated flowers veined on deep crimson purple (W.A.B.)

Blue Veins Fringed: deeply fringed flowers of various shades of purple, rose and lavender, penciled and veined with deep bluish purple (W.A.B.)

Prince of Wurtemburg: extra large flowers; dark purple with tigered throat (W.A.B.)

Princess of Wurtemburg: flowers rose color with large white throat veined with maroon (W.A.B.)

Superbissima: enormous flowers, purple and crimson, very clear throat, richly veined (W.A.B.)

White-throated Grandiflora: dark blood-red with pure white throat (W.A.B.)

Yellow-throated: yellow throat, veined like Salpiglossis (W.A.B.)

1895–1902

Belle Etoile: small-flowered, single; white and purple (D.L.)

Brilliant Rose: single flowers, rose colored (D.L.)

California Giants: produced in California; flowers 4 in. in diameter with a great range of colors, presenting a wonderful combination of stripings, veinings, and blotchings; single (D.L., T.W.W.)

1903–1914
Snowball: dwarf, bushy, sweet scented; flowers white (T.W.W.)

DRUMMONDS PHLOX

Phlox drummondii   Polemoniaceae

DESCRIPTION FROM HORTUS THIRD

Phlox Drummondii—flowers rose-red varying to white, buff, pink, red, and purple, to 1 in. across in dense cymose clusters. South Central Texas.
Color forms represented by such cvs. as —'Alba Oculata,' 'Atropurpurea,' 'Caerulea-striata,' 'Carnea,' 'Coccinea,' 'Isabellina,' 'Rosea,' 'Violacea.'
cv. 'Gigantea': large-flowered strain with a wide range of colors.
cv. 'Grandiflora'—florets purple, white beneath.
cv. 'Rotundata'—corolla lobes broad.
cv. 'Twinkle'—Star Phlox, a group with narrow, cuspidate corolla lobes, which are often cut and fringed; also known as cv. 'Sternenzauber.'
Other cvs.—'Leopoldii' and 'Nana Compacta.'

1865–1868

Phlox drummondii alba: pure white (J.B.)

drummondii azurea, or Large Blue: white eye; the nearest to blue of the phloxes, but really a fine purple (J.B.)

Chamois Rose: delicate and fine; new (J.B., J.V.)

P. drummondii coccinea, or Brilliant Scarlet: pure deep scarlet (J.B., J.V.)

drummondii flore-albo: pure white (J.V.)
—*occulata*: pure white, with purple eye (J.V.)

Isabellina: light, dull yellow; new (J.B.)

P. drummondii Leopoldi: deep pink, white eye (J.V., J.B.)

drummondii marmorata violacea: violet, marbled; new (J.V., J.B.)

Napoleon: purplish crimson with black eye (J.B.)

Princess Royal: purple, white striped (J.B.)

Queen Victoria: violet with white eye (J.B.)

P. drummondii Radowitzii: rose striped with white (J.B., J.V.)

—*Kermesina striata*: crimson, striped with white; new
—*violacea*: violet, striped with white (J.V.)

drummondii rosea marmorata: rose, marbled; white eye (J.V.)

drummondii variabilis: violet and lilac (J.V.)

Violet Queen (may be the same as Queen Victoria): violet, with large white eye (J.V.)

William I: crimson with white stripe (J.B.)

1869–1873

P. drummondii atropurpurea striata: deep purple and white striped (J.B.)

> *drummondii Graf Gero*: dwarf, pyramidal, floriferous; flowers alternately red and white with white eye; 8 in. high (J.B.)
>
> *drummondii Heynholdi*: true scarlet with a slight tinge of copper; dwarf; introduced from France (J.B.)

James Vick claimed: "I grow my own seed of the Phlox exclusively, devoting much time and means to its improvement, and have no hesitation in saying my Phlox seed is the best the world produces." (Vick, *Floral Guide*, 1873)

1874–1877

P. drummondii Bedmani: salmon tint, large and distinct; dwarf (P.H.)

> *drummondii grandiflora*: improved, with flowers unusually large, round, and of great substance; a new class (J.V.)
> —*splendens*: "New variety from Europe. Large bright scarlet flowers with a conspicuous white eye, the center of which is encircled with a well defined violet edge." (Henderson, 1875)

drummondii Heynholdii Cardinalis: "copper rose tinge, as well as paleness of the undersides of the petals in the original have given way to the pure intense fiery scarlet." (Henderson, 1874)

1878–1883

Black Warrior: very dark purple (W.A.B., D.L.)

Carmine Queen: large white eye (W.A.B.)

Empress Eugenie: rose, marbled (W.A.B.)

P. drummondii Hortensiaeflora alba: dwarf with large white flowers, vigorous and profuse blooming, withstands unfavorable weather. (J.B.)

> *drummondii nana compacta*: dwarf class with smaller flower trusses than above.
> —Chamois Rose (W.A.B.) (see 1865–1868; not advertised as dwarf)
> —Fireball: scarlet (W.A.B.)

—Snowball: white (W.A.B.)
Vermillion (W.A.B., no description)

1884–1895

P. drummondii cuspidata: "Each lobe of the corolla is furnished with one long, narrow, pointed segment and two lesser ones, and all bordered with a narrow, white band." (Charles S. Sargent, *Garden and Forest*, 1888) Evolved from *P. drummondii grandiflora fimbriata*.
—Star of Quedlenburg: dark violet-purple (J.B., J.T.)

drummondii flore pleno, or Double Phlox: "Florists have doubled the flowers of a white and a red variety, but the doubling is only semi-double and (in regards to beauty of the flowers) is more of an injury than a benefit. Double white is fairly true from seed; only small percentage of double red" (Sargent, *Garden and Forest*, 1888)
—*alba*, or Double White (W.A.B.)
—*coccinea*: or Double Scarlet (W.A.B.)
—mixed (D.L.)

drummondii nana compacta: mixed (D.L.)
—Cinnabarina: vermillion-scarlet (W.A.B.)
—*coccinea striata*: crimson flowers striped white; similar to Fireball (W.A.B.)
—Victoria: crimson-scarlet (W.A.B.)

1896–1914

Three major classes were sold in American catalogues: *P. drummondii cuspidata* (Star Phlox), *P. drummondii grandiflora* (Large flowering Phlox), and *P. drummondii nana compacta* (Dwarf Phlox). See, for example, the 1908 list from James Vick's Sons.

PORTULACA OR MOSS ROSE

Portulaca grandiflora   Portulacaceae

DESCRIPTION FROM HORTUS THIRD

Portulaca grandiflora—prostrate, or to 1 foot high; flowers rose, red, yellow, white, often striped; 1 in. across or more. Brazil, Argentina, Uruguay.

1865–1870

Portulaca alba: white (J.B.)
—*striata*: white striped with rose (J.V.)

aurea: straw-colored (J.V., J.B.)
—*vera*: deep golden yellow (J.V.)
—*striata*: sulphur yellow, striped with gold (J.V.)

Blensonii: light scarlet (J.V.)

caryophylloides: rose striped with deep carmine (J.V.); carnation striped, white and crimson (J.B.)

New Rose or *P. rosea*: rose (J.B., J.V.)

Rose-flowered, Double: "a perfectly double variety, as much so as the most perfect Rose, and of many brilliant colors as well as striped. . . . About ¾ of all plants produced from seed are double. Package of 25 seeds, first quality, imported—75¢ [compared to 10¢ per package for other portulaca seed]" (Vick, 1865). Also offered in separate colors: crimson, rosy-purple, white, white striped with red, orange, yellow.

P. splendens: crimson purple (J.B.)

Striped: red and white (J.B.)

P. Thellusonii: fine crimson (J.V.); splendid scarlet (J.B.).

Thoroburnii: deep orange (J.B.).

1871–1880
P. alba, *P. aurantiaca*, *P. flore pleno* (rare seed), *P. splendens*, *P. Thellusoni* (D.L., J.B., J.V.)

P. Bedmani: large pure white flowers with reddish purple eye, dark red stalk and branches (P.H.)

Golden Striped: straw colored with golden orange stripes (P.H.)

New Pink: large flowers, delicate pink shade (P.H.)

1881–1900
Portulaca Double Rose-flowered Mix (W.A.B.)

    Splendid Single Mix (W.A.B.)

James Thorburn's Portulaca List of 1890:
    alba oculata: white and purple
    albiflora: white
    aurantiaca: orange
    Bronze
    Large-flowered: flesh, rose, salmon, buff, striped, double scarlet, orange, carnation striped, salmon, white, purple, mix
    Light Straw: yellow
    Light Rose: light rose
    Pheasant's Eye: red and white
    Red Carnation Striped
    rosea: rose
    splendens: purple

sulphurea: sulphur
Thorburni: [not described]

1901–1914

P. splendens aurea: yellow (T.B.)

MIGNONETTE

Reseda odorata   Resedaceae

DESCRIPTION FROM HORTUS THIRD

Reseda odorata—becoming decumbent; racemes dense; flowers yellowish-white, very fragrant, sepals and petals 6, anthers orange. N. Africa. Cultivated abroad for essential oil used in perfumery.
cv. 'Grandiflora'—a large-flowered garden form.

1865–1880

Reseda odorata arborea or Tree Mignonette: "more stocky and tree-like; excellent for the conservatory, but where Mignonette is treated as an annual in the garden, no better than the common variety." (Vick, 1865, p. 27)

    odorata grandiflora: orange and buff (J.B., D.L.); large-flowered . . . but no better for ordinary purposes. (Vick, 1865)
    —*ameliorata*: large-flowered, red (J.B., J.V.)

Parson's White: robust flowers larger and showing more white than the common sort, improved (J.V.)

1880–1886

Crimson Giant: large flowers tinged crimson (W.A.B.)

Golden Queen: flowers golden; dense pyramidal habit (J.B., W.A.B.)

Machet: "Of French origin, and pronounced the best for pot culture; comes perfectly true from seed. The dwarf and vigorous plants are of pyramidal growth, with very thick, dark-green leaves; they throw up numerous stout flower stalks, terminated by long and broad spikes of deliciously scented red flowers." (Burpee, "Novelties," 1884)

Mile's Hybrid Spiral: flowers cream and buff; robust and free flowering (J.B., W.A.B., J.H.)

New Giant: large flowers; "best variety for beekeeping" (Breck, 1886)

The Prize, or White Prize: white (J.B.)

Pyramidal Large-flowered: flowers orange-red; pyramidal habit (J.B.)

1887–1900

Allen's Defiance: red and white; large spikes; florist's favorite (J.B., D.L.)

Breck's Giant Machet: red and white (J.B.)

Gabrielle: red (J.B., W.A.B.)

Golden Machet: golden tinged flowers; good for pot culture (J.B.)

Fordhook's Finest: selected forms (W.A.B.)

Pyramidal Dwarf Bush, or —*pumila compacta*: dwarf habit (J.T.)

Quaker City: "After growing half an acre in this grand new Mignonette, we feel we cannot say too much in its praise. . . . The flower spikes are of great substance, very full, rounded at the top, of a handsome gold-red color, and unusually rich, sweet fragrance. The foliage is very dark-green, and the magnificent heads are produced profusely and continuously. It should be in every garden, and is also specially adapted for pot culture and florists' use." (Burpee, "Flower Seed Novelties," *Farm Annual*, 1888, p. 91)

Upright: tall growing (J.T.)

Victoria: red (J.B.)

1901–1914

Bird's Mammoth: giant red flower spikes; plants pyramidal in form; fine for pots (D.L.)

Bismarck: Machet type, reddish-colored flowers (J.B.)

MARIGOLD

Tagetes sp.　　　　　　　　　　　　　　　　Compositae

DESCRIPTIONS FROM HORTUS THIRD

Tagetes erecta—African or Aztec Marigold. stout, erect annuals to 3 ft.; leaves pinnate; ray flowers 5–8 in the wild form, many and often 2-lipped or quilled in cvs., light yellow to orange. Mexico, Central America; now naturalized in many warm regions.

T. patula—French Marigold. bushy, ½–1½ ft.; ray florets few to many, yellow, orange, red-brown or particolored. Mexico, Guatemala.

T. signata, now *T. tenuifolia*—slender annual, to 2 ft.; ray florets few, ½ in., yellow, elongated. Mexico, Central America.

"Pumila" is a group name covering various cvs. of low, dwarf, compact habit, 1 ft. or less.

1865–1885

Tagetes erecta: generally offered in orange and lemon shades (J.B., P.H., D.L., J.V.); Vick offered African marigolds in tall and dwarf forms—Tall Orange (double), Tall Sulphur, Tall Quilled (orange, double), Tall Quilled Sulphur, Dwarf mixed

patula: mixed colors (J.B., P.H., D.L., J.V.); Vick offered Tall Orange, Tall Brown, Tall Striped, Dwarf Sulphur, Dwarf Brown, Striped Dwarf, Dunnett's New Orange (very superior), and — *aurea nana flore pleno* (pure orange dwarf French, very double)

signata: "A novelty of 1863, and a most beautiful plant . . . from 12 to 18 in. in height. . . . They are beautiful as single plants, and form a delightful bed on the lawn." (Vick, 1865)
—*pumila* form also sold by J.B., P.H., D.L.

1886–1899

T. erecta Eldorado: large, bushy plants, enormous flowers (3½ to four in. across), in four shades—light primrose, lemon, rich golden yellow, and deep orange, imbricated (J.B., W.A.B., J.T., J.V.)

patula Superb Striped: Scotch prize strain; mixed colors (W.A.B.); Purple and Gold: striped (W.A.B.)

1900–1914

T. erecta Delight of Garden: dwarf, double, rich yellow (D.L.)
Lemon Queen: tall, lemon yellow (D.L., J.B., J.V.)
Nugget of Gold: golden yellow, double (D.L.)
Prince of Orange (J.V.)
Pride of the Garden: double, golden yellow (J.B., D.L., J.V.)
Quilled: mixed colors (J.B.)

patula Dwarf Brownee, Little Brownie, or Legion of Honor: golden yellow marked velvety red or maroon, single (J.B., D.L., J.V.)
Gold Margin or Gold Edged: velvety maroon, margined with gold (J.B., J.V.)

signata pumila Golden Ring: brown and yellow (J.B.) Cloth of Gold: yellow (J.B.)

NASTURTIUM

Tropaeolum sp.   Tropaeolaceae

DESCRIPTIONS FROM HORTUS THIRD

Tropaeolum majus—Tall Nasturtium, usually climbing, somewhat succulent; leaves orbicular; flowers 2½ in. across, of various shades of yellow, orange, or red, sometimes striped and spotted. Andean South America. Double-flowered and dwarf cvs. offered. cv. 'Burpeei'—flowers very double, not producing seeds, propagates vegetatively.

T. minus—Dwarf Nasturtium, not climbing, but more or less scrambling; leaves with each vein ending in a point; flowers 1½ in. across or less, the lower 3 petals with a dark central spot. Andean South America.

T. Lobbianum, now *T. peltophorum*—stems climbing, hairy; leaves orbicular, long-petioled, hairy beneath; flowers 1 in. long, orange-red, the 3 lower petals long-clawed, margins toothed. Andean South America.

T. peregrinum—Canary-Bird Nasturtium, stems climbing; leaves deeply 5-lobed, flowers to 1 in. across, canary-yellow, the upper petals erect and fimbriate, the spur curved, green. Probably Andes of Peru and Ecuador.

1865–1870

Tropaeolum Lobbianum: summer beds and greenhouse. (The following selection offered in Breck's 1868 and Vick's 1865 lists.)

  Caroline Schmidt: deep scarlet
  Duc de Luynes: very dark crimson
  Duc de Malakoff: straw color, edged rose and spotted red
  Duc de Vicenza: pale lemon with vermilion spots
  flamula grandiflora: yellow streaked with carmine
  Garibaldi: orange shaded with scarlet
  Geant des Battailles: carmine
  Monsieur Calmet: lemon spotted with crimson
  Napoleon III: yellow striped with vermilion
  Prince Imperial: ruby spotted with maroon
  Queen Victoria: vermilion and scarlet striped
  Roi des Noires: almost black
  Smith, Lilli: orange-scarlet

T. major (Breck and Vick lists)

atropurpureum or *atrosanguinea*: dark crimson
carnea: bluish
coccineum: scarlet
Dunnett's Orange: dark orange
Edward Otto: bronze, silky, new
luteum: yellow
Scheuerianum: straw or cream, spotted with crimson
—*coccineum*: scarlet, striped
Shillingii: spotted
Schulzii: brilliant scarlet
Common Mixed: seed pods when green used for pickles

Tom Thumb (Breck and Vick lists); some firms referred to this class as a separate species, *T. nanum*; others relegated the Tom Thumbs to a race of *T. minus*, dwarf nasturtiums.

Beauty: yellow blotched with red
Carter's: scarlet
Crimson: dark crimson
Crystal Palace Gem: sulphur spotted with maroon
Spotted: bright yellow spotted crimson
Yellow: pure yellow

T. peregrinum: fine climber, with an abundance of yellow flowers all the summer and autumn; fine for arbors, trellises, etc. (J.V.)

1871–1878

T. Lobbianum, Triomphe de Gand: orange scarlet (P.H.)

Tom Thumb, King of Tom Thumb: deep scarlet (J.B., J.V.)
King Theodore: flowers very dark, new (J.B., J.V.)
Pearl: nearly white (J.B.)
Rose: rose color, new (J.V.)
Ruby King: carmine (J.B.)

1879–1885

Burpee listed all Tom Thumb forms listed above and, in addition,
Golden King: bright gold Tom Thumb

Landreth's offered *T. nanum* in mixed colors; *T. majus atropurpurea, coccinea, luteum.*

1886–1890

Lobb's Nasturtium, Spitfire: scarlet (J.H.)

Tom Thumb, Chameleon: flowers crimson, bronze, and gold bordered and flamed, changing hues almost daily; compact and graceful, profuse bloomer; excellent pot plant. (W.A.B.)
Cloth of Gold: clear bright yellow foliage; deep scarlet, flowers; dwarf and compact; produces very little seed (W.A.B., J.T.)
Coeruleum-roseum: peach color (W.A.B.)

Empress of India: awarded 1st Class Certificate by R.H.S.; dwarf busy habit; leaves dark purplish-blue color; flowers crimson-scarlet (W.A.B., J.T.)

1891–1898

Lobb's, Prince Bismarck: new, no description (T.W.W.)

Tom Thumb, Aurora: brown-red (W.A.B., J.B.)
Breck's Rainbow: mixed colors (J.B.)
Gen. Jacqueminot: intense deep red (W.A.B.)
Cattell's Crimson: showy (J.T., D.L.)
Prince Henry: light yellow marbled scarlet (W.A.B.)

1900–1908

Lobb's, Asa Gray: yellowish white (J.B., J.T.)
Crown Prince of Prussia; blood-red (J.B., J.T.)
Lucifer: very dark scarlet, dark foliage (J.B., J.T.)

T. majus Heinemanni: chocolate (J.B., J.T.)
hemisphaericum: orange (J.B., J.T.)
Vesuvius: salmon (D.L.)
Von Moltke: bluish rose (D.L.)

Tom Thumb, Bronze Curled: flowers of bronze metallic lustre (D.L.)
Ivy-leaved: dark colored foliage and flowers of a new form; desirable for use in window boxes (D.L.)
Tom Pouce: smaller leaves, good for baskets and window boxes, mixed colors (D.L.)

PANSY

Viola tricolor   Violaceae

DESCRIPTION FROM HORTUS THIRD

Viola tricolor, Garden Pansy now *V.* × *Wittrockiana* —Hybrid between *V. tricolor* and apparently *V. lutea*, together with *V. altaica*. Stems leafy; flowers 2–5 in. across, rounded in outline, flattened, variously colored.

1865–1870

Pansies were known by their "fancy" names at this time and are listed accordingly below.

English Prize: seed from choicest English prize flowers (J.V.)

Faust, or King of the Blacks: almost coal black, coming true from seed (P.H., J.V.)

Marbled Purple: new colors, marbled (J.V.)

Odier, or 5-Blotched: large blotches in center (J.B.)

Pure Yellow: always true to color (J.V., J.B.)

Red: very bright (J.V.)

Sky Blue: shades of light blue (J.V., J.B.)

Striped and Mottled: very showy, various colors (J.V.)

White: sometimes slightly marked with red or purple (J.V.)

Yellow Margined: margin or belt of yellow, new (J.V.)

1871–1879

Cliveden Purple: rich, deep purple (J.V.)

Emperor William: received from Germany, new; large flowers of ultramarine blue with purple-violet eye (J.V., P.H.)

Mahogany-Colored: very fine (J.V.)

Violet: with white border (J.B., J.V., P.H.)

White Treasure: pure white finely marked with yellow eye; flowers have a transparent tinge (P.H.)

1880–1885

Belgian Striped and Mottled: various colors (W.A.B.)

Bronze Colored: self-colored (W.A.B.)

Brownish Red: (W.A.B.)

Fawn-Color: (W.A.B.)

Havana-Brown: new shades (W.A.B.)

Lord Beaconsfield: "The ground color of the flower of this elegant Pansy is purple-violet, shading off on the top petals only to a whitish hue; a peculiarity which lends to the whole flower an unusually bright appearance." (Breck, *Catalogue of Novelties*, 1883)

Snow Queen: "The flowers of this charming sort differ entirely from the ordinary white Pansy. They are of a particularly delicate satiny white with a slight tinge of yellow towards the centre. It reproduces itself exactly from seed, and comes highly recommended." (Breck, *Catalogue of Novelties*, 1883); (W.A.B.)

Superb: extra large flowers, richly varied colors; diameter of a single flower twice as large as a trade dollar (W.A.B.)

1886–1893

Azure blue: very fine (similar to Sky Blue and Light Blue) (J.B., W.A.B.)

Bronze (J.B., P.H.)

Cassier's Super Pansy

Gold Margined (J.B., W.A.B.)

Maxima quadricolor: blending of peculiar colors (J.B.)

Red Riding Hood: entirely new color of Imperial Giant strain; circular form; solid red (W.A.B., J.V.)

1894–1905

Bugnot's Superb German: largest single flowers, striped and blotched (W.A.B., J.H., J.T.)

Giant Five-Spotted: large dark blotch on each petal, most flowers have a margin of white or yellow (D.L., J.T.)

Landreth's Philadelphia: special strain; flowers 2½ in. in diameter (D.L.)

Masterpiece: highest perfection of the 5-spotted pansies (D.L.)

Peacock: blue (D.L.)

Prince Bismarck: golden bronze (D.L.)

Victoria Red: claret-red (W.A.B., J.B., D.L.)

1906–1914

(The following giant-flowering sorts are from James Vick's Sons, 1908)

Adonis: light blue

Andromeda: delicate apple-blossom, showing a soft lavender or lilac hue, rendered more effective by a darker lining

Bridesmaid: rosy white ground, blotched; new

Emperor Franz Joseph: white with large violet blotches; a superior bedding variety

Fire King: golden yellow, upper petals purple

Freya: dark purple, with broad pure white margin

Golden Queen: pure yellow

Madam Perret: frilled petals; dark wine, pink, and red shades beautifully veined, all with white margin

Prince Henry: darkest blue

Psyche: curled and undulated

Ruby King: superb red shades

Snowflake: extra pure white (also D.L.)

(Standard sorts included the following)

Cardinal: dark red

Diana: pale yellow or cream; flowers slightly ruffled (D.L.)

Red Riding Hood: brilliant red (W.A.B., D.L.)

ZINNIA

Zinnia sp.   Compositae

DESCRIPTIONS FROM HORTUS THIRD

Zinnia elegans—disc flowers yellow to purple when present, ray florets 8–20, spatulate, and usually red in the wild type, but in the cvs. often several times as many and elongated or broader, often twisted or tubular, and of every color but blue. Mexico. There are many races and cvs. available, differing in stature, size of heads, length of peduncles, and color, form, and number of ray florets.
cvs.—'Gracillima,' 'Pumila,' 'Scabiosiflora,' 'Striate' among others.

Z. Haageana—disc florets orange, ray florets 8–9 and orange in the wild type, but more numerous and usually bicolored red-and-yellow or -orange in cvs. Mexico.
cvs.—'Old Mexico' and 'Persian Carpet.'

Z. mexicana, now *Z. Haageana*

Z. multiflora, *Z. pauciflora*, and *Z. tenuiflora* now *Z. peruviana*—differs from *Z. Haageana* in its yellow to scarlet ray florets, rounded, often red-tipped receptacle scales, and linear-oblanceolate disc achenes with a single awn. Arizona and Mexico to Argentina; also West Indies.

"New Double-Flowered Zinnia: By favor of M. Vilmorin, of Paris, has favored us with an engraving of the new candidate. . . . In the present improvement the petals of the central flowers have . . . been developed into the ligulate shape, and the result is, as in the Dahlia, what we call a double flower. Though it is the same species as the one in cultivation, M. Vilmorin says he has never succeeded in obtaining seedlings with more than two rows of petals. He received the first seeds two years ago from M. Grazan, gardener, at Bagneres, who had them sent from India." (Thomas Meehan in *The Gardener's Monthly*, January 1861)

1865–1870

Zinnia elegans in separate colors below (J.B.)
—*alba*: white
—*aurea*: deep gold
—*coccinea*; scarlet
—*purpurea*: purple
—*sulphurea*: yellow

New Double Flowered: most important acquisition in many years (J.B., J.V.)

Z. mexicana: dwarf species with yellow flowers flushed with orange (J.B.)

tagetiflora flore pleno,: like a marigold (J.B.)

pumila or *elegans pumila*: dwarf (J.B., J.V.)

1871–1878

Peter Henderson offered *Z. elegans*, —*alba*, *Z. Haageana flore pleno*, *Z. elegans pumila alba pleno samonea* (new dwarf), —*coccinea*, and *violaceae*.

Z. Haageana flore-pleno: "Messrs. Haage and Schmidt . . . regard [this flower] as the best novelty amongst annuals which the season has produced. The typical form of the species . . . is pretty freely cultivated as a neat, low-growing, tufted, yellow flowered annual, and has long since become a favorite from its usefulness for dried or winter bouquets. This double-flowered variety differs in the form and doubleness of the flower-heads, which are perfectly rosette shaped, and, moreover, is reported to be constant from seeds. The habit is the same as that of the single flowered sort. . . . The florets are densely imbricated. . . . Their color is a deep orange-yellow, keeping its lustre when dried. . . . It is likely to be a good bedding plant, blooming continuously till frost sets in" (Thomas Meehan in *The Gardener's Monthly*, January 1872, p. 29). (J.B., P.H., D.L., J.V.)

1879–1887

Burpee offered Zinnia Double choice mix and Pure White.

Z. hybrida: in mixed colors, 2 feet [presumably *Z. elegans*] (D.L.)

1888–1895

Burpee's Superb Double: "A selection made for us by a special grower, whose strain is unrivaled" (W.A.B., 1888)

Double White (D.L., J.T.)

Large-Flowered Dwarf: compact, bushy,—*pumila* types. Also known as Tom Thumb. (W.A.B., J.T., T.W.W.)

Mammoth, New: "A particularly fine new class . . . differing from the older ones from its unusually robust habit of growth and the immense size (five to six inches across) of its perfectly-formed, very double flowers . . . [which] are uninfluenced by heat and remain in good condition for several weeks." (W.A.B., 1888)

Salmon Rose: double form (J.T.)

Striped Zebra: "The petals are as distinctly striped as the most beautiful Carnations or Balsams. In some, the flowers are yellow, splashed crimson or maroon, white with rose or lilac, and many other exquisite tints of coloring hitherto unknown in Zinnias. Flowers from this strain were exhibited before the Committee of the Royal Horticultural Society in August, 1886, and were unanimously awarded a First Class Certificate" (W.A.B., 1888). (Offered by J.T. as Double Striped)

1896–1905

Breck's Defiance: in separate colors and mixed—carmine crimson, flesh pink, golden yellow, light yellow, lilac, purple, rose, scarlet (light and dark), striped, white. "The plants are of vigorous branching habit and produce freely immense double flowers in an endless variety of splendid colors; not the old hard red but innumerable delicate and velvety shades. The separate colors are saved from the same collection which is admitted to be the finest in Europe." (Breck, 1900)

Curled and Crested: new class of plumed and twisted petals (J.B., D.L.)

Lilliput: double flowers in mixed colors, 9 inches high (J.B.)

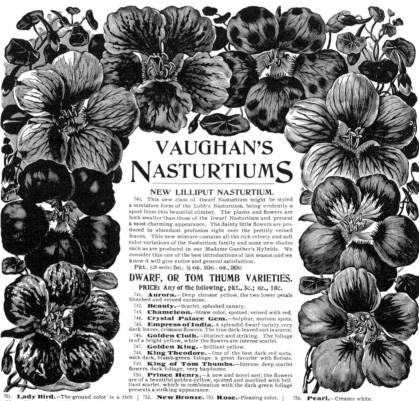

VAUGHAN'S NASTURTIUMS

NEW LILLIPUT NASTURTIUM.

740. This new class of Dwarf Nasturtium might be styled a miniature form of the Lobb's Nasturtium, being evidently a sport from this beautiful climber. The plants and flowers are both smaller than those of the Dwarf Nasturtium and present a most charming appearance. The dainty little flowers are produced in abundant profusion right over the prettily veined leaves. This new mixture contains all the rich velvety and soft color variations of the Nasturtium family and some new shades such as are produced in our Madame Gunther's Hybrids. We consider this one of the best introductions of last season and we know it will give entire and general satisfaction.

Pkt., (25 seds) 5c.; ½ oz., 10c.; oz., 20c.

DWARF, OR TOM THUMB VARIETIES.

PRICE: Any of the following, pkt., 3c.; oz., 10c.

741. **Aurora.**—Deep chrome yellow, the two lower petals blotched and veined carmine.
742. **Beauty.**—Scarlet, splashed canary.
743. **Chameleon.**—Straw color, spotted; veined with red.
744. **Crystal Palace Gem.**—Sulphur, maroon spots.
745. **Empress of India.**—A splendid dwarf variety, very dark leaves; crimson flowers. The true dark-leaved sort is scarce.
746. **Golden Cloth.**—Distinct and striking. The foliage is of a bright yellow, while the flowers are intense scarlet.
747. **Golden King.**—Brilliant yellow.
748. **King Theodore.**—One of the best dark red sorts, with dark, bluish-green foliage; a great favorite with florists.
749. **King of Tom Thumbs.**—Intense deep scarlet flowers, dark foliage; very handsome.
750. **Prince Henry.**—A new and novel sort; the flowers are of a beautiful golden-yellow, spotted and marbled with brilliant scarlet, which in combination with the dark green foliage presents a striking appearance.

751. **Lady Bird.**—The ground color is a rich golden-yellow, each petal is barred with a broad vein of bright, ruby crimson.
752. **New Bronze.**
753. **Rose.**—Pleasing color.
754. **Ruby King.**—Peculiar blue-tinted red.
755. **Scarlet King.**—Brilliant scarlet.
756. **Pearl.**—Creamy white.
757. **Spotted King.**—A very handsome sort.
758. **Terra Cotta.**—A novel color.

NOTES

INTRODUCTION

1. Charles Van Ravenswaay, *A Nineteenth-Century Garden* (New York, 1977), 8.

CHAPTER I

1. Leslie R. Hawthorn and Leonard H. Pollard, *Vegetable and Flower Seed Production* (New York, 1954), 22, 32.
2. John Harvey, *Early Nurserymen* (London, 1974), 14.
3. Burnet Landreth, Jr., "Address before the Poor Richard Club of Philadelphia on the 150th Anniversary of the David Landreth Seed Co." (1934), 5. Pennsylvania Collection, Pennsylvania Horticultural Society Library, Philadelphia.
4. James Vick, *Illustrated Catalogue of Seeds* (1865), 2.
5. David Landreth, *Landreth's Rural Register and Almanac* (Philadelphia, 1886), 2.
6. Ibid.
7. [Thomas Meehan], "Books, Catalogues, &c. Review," *The Gardener's Monthly and Horticultural Advertiser* (June 1872), 181.
8. James Vick, "Vick's Celebrated Flower-Farm," *Vick's Floral Guide* (1875, no. 2), 63.
9. Ibid.
10. Ibid.
11. Samuel P. Hays, *The Response to Industrialism 1885–1914* (Chicago, 1957), 6.
12. Peter Henderson, *Practical Floriculture*, new and enlarged ed. (New York, 1874), 186–88.
13. Hays, *Industrialism*, 8.
14. Daniel Walker Howe, "Victorian Culture in America," in *Victorian America* (Philadelphia, 1976), 17.
15. Hays, *Industrialism*, 8.

16. Ibid.
17. Howe, "Victorian Culture," 4.
18. Ibid., 9–11.
19. Ibid., 19.
20. [Thomas Meehan], "The Floral Sabbath," *The Gardener's Monthly* (September 1872), 269.
21. Nathaniel H. Egleston, *Villages and Village Life* (New York, 1878), 99.
22. Howe, "Victorian Culture," 26.
23. Egleston, *Village Life*, 97.
24. [Thomas Meehan], "Hints for January," *The Gardener's Monthly* (January 1872), 1–2.
25. Hugo de Vries, *Plant Breeding* (Chicago, 1907), 1–6.
26. Liberty Hyde Bailey, *The Survival of the Unlike*, 4th ed. (New York, 1901), 178.
27. Richard Gorer, *The Development of Garden Flowers* (London, 1970), 17.
28. Bailey, *Survival*, 178.
29. H. F. Roberts, *Plant Hybridization Before Mendel* (New York, 1965), 286–90.
30. S. L. Emsweller et al., "Improvement of Flowers by Breeding," *Yearbook of Agriculture, 1937* (Washington, D.C., 1937), 891.
31. Bailey, *Survival*, 218.

CHAPTER II

1. John Harvey, *Early Gardening Catalogues* (London, 1972), 152.
2. Ibid.
3. Ibid., 127, 129, 131.
4. Peter Henderson, *Gardening for Pleasure* (New York, 1891), 27.
5. Bernard M'Mahon, *American Gardener's Calendar*, 11th ed. (Philadelphia, 1857), vii.
6. Pennsylvania Horticultural Society, *From Seed to Flower* (Philadelphia, 1976), 95.
7. M'Mahon, *Calendar*, 2.
8. William N. White, *Gardening for the South* (New York, 1858), 9.
9. T. W. Wood & Sons, *Catalogue for 1894* (Richmond, Virginia, 1894).
10. D. W. Beadle, *Canadian Fruit, Flower, and Kitchen Garden* (Toronto, 1872), 346.
11. Van Ravenswaay, *Nineteenth-Century Garden*, 7.
12. M'Mahon, *Calendar*, 169–70.
13. William Cobbett, *The American Gardener* (London, 1821), paragraphs 375, 339.

14. M'Mahon, *Calendar*, 341.
15. White, *South*, 256.
16. Vick, *Floral Guide* (1877, no. 1), 1.
17. Peter Henderson, *Henderson's Handbook of Plants* (New York, 1881), 24.
18. R. & J. Farquhar, Seedsmen, *Gardening in a Nutshell* (Boston, 1898).
19. In 1952 the International Botanical Congress Committee on Horticultural Nomenclature and Registration, at the Thirteenth International Horticultural Congress, in London, formally initiated specific rules designed to achieve a concise and standardized registration of new cultivars through strict adherence to specified regulations. These rules, which have undergone minor revisions in 1957, 1960, and 1969, are stated in the *International Code of Nomenclature of Cultivated Plants*, J.S.L. Gilmour et al., eds. (Utrecht: International Bureau for Plant Taxonomy and Nomenclature, 1969).
20. Bailey, *Survival*, 358.
21. Van Ravenswaay, *Nineteenth-Century Garden*, 8.
22. Geoffrey Taylor, *The Victorian Flower Garden* (Essex, 1952), 73.
23. Ibid.
24. Ibid., 74.
25. Mariana Griswold Van Rensselaer, *Art Out-of-Doors* (New York, 1925), 141.
26. Jane Loudon, "Sowing Annuals," *Ladies Magazine of Gardening* (January 1842), 106.
27. [Thomas Meehan], "Hints for May," *The Gardener's Monthly* (May 1861), 129.
28. Henderson, *Practical Floriculture*, 26.
29. Ibid.
30. "Letter from Paris Correspondent," *Gardener's Monthly* (July 1861), 223.
31. R. Morris Copeland, *Country Life: A Handbook of Agriculture, Horticulture, and Landscape Gardening* (Boston, 1859), 547.
32. Van Rensselaer, *Art Out-of-Doors*, 142.
33. [Thomas Meehan], "Hints for March," *Gardener's Monthly* (March 1872), 1.
34. [Thomas Meehan], "Books, Catalogues, &c. Review," *Gardener's Monthly* (January 1872), 1.
35. William Robinson, *The English Flower Garden*, 7th ed. (London, 1901), 111.
36. [William Robinson], "Flower Garden Notes," *The Garden* 39 (January 31, 1891): 95–96.
37. Liberty Hyde Bailey, "Use of Wild Flowers in Cultivation," *The House Beautiful* (June 1902), 51.
38. Van Rensselaer, *Art Out-of-Doors*, 160.

39. Louis Beebe Wilder, *Colour in My Garden* (New York, 1918), 287.
40. Gertrude Jekyll, *Colour in the Flower Garden* (London, 1908), vi.
41. Louisa Yeomans King, *The Well-Considered Garden* (New York, 1915), 221.

CHAPTER III

1. J. Fraser and A. Hemsley, eds., *Johnson's Gardener's Dictionary and Cultural Instructor* (New York, 1917), 216, 381.
2. Ibid., 230–31, 339–40.
3. Ibid., 877.
4. John Gerard, *The Herbal or General History of Plants* (1633; reprint, New York: Dover Publications, 1975), 251–52.
5. John Parkinson, *A Garden of Pleasant Flowers; Paradisi in Sole Paradisus Terrestris* (1629; reprint, New York: Dover Publications, 1976), 280–81.
6. Fraser, *Johnson's Dictionary*, 877.
7. Ibid.
8. Emsweller et al., "Improvement of Flowers," 926.
9. Buckner Hollingsworth, *Flower Chronicles* (New Brunswick, 1958), 224.
10. [Charles M. Hovey], "Floricultural Notes," *The Magazine of Horticulture* (September 1866), 371.
11. [William Robinson], in *The Garden* 2 (September 21, 1872), 249.
12. Fraser, *Johnson's Dictionary*, 188.
13. Joseph Paxton, *Paxton's Botanical Dictionary* (London, 1868), 119.
14. *Hortus Third* (New York, 1976), 241–42.
15. Vick, *Floral Guide* (1875, no. 1), 24.
16. [Charles M. Hovey], "New Feathered Crimson Celosia," *The Magazine of Horticulture* (January 1868), 21.
17. Vick, *Floral Guide* (1875, no. 1), 24.
18. Vick, *Floral Guide* (1877, no. 1), 23.
19. Fraser, *Johnson's Dictionary*, 844.
20. Ibid.
21. Roy Genders, *The Cottage Garden* (London, 1969), 320.
22. Parkinson, *Paradisi in Sole*, 303.
23. Genders, *Cottage Garden*, 320.
24. Ann Leighton, *American Gardens in the Eighteenth Century* (Boston, 1976), 455.
25. Beadle, *Canadian Garden*, 348–49.
26. Vick, *Illustrated Catalogue and Floral Guide* (Spring 1865), 26.
27. Fraser, *Johnson's Dictionary*, 453.
28. Leighton, *American Gardens*, 397.

29. Beadle, *Canadian Garden*, 347.
30. Genders, *Cottage Garden*, 291.
31. Ibid.
32. Philip Miller, *The Gardeners Dictionary*, abridged ed. (1754; reprint, Codicote, Hitchin, Eng.: Wheldon and Wesley, 1969), 1197–98.
33. Genders, *Cottage Garden*, 291.
34. Beadle, *Canadian Garden*, 249.
35. Emsweller et al., "Improvement of Flowers," 926.
36. Claire Shaver Haughton, *Green Immigrants* (New York, 1978), 27.
37. Emsweller et al., "Improvement of Flowers," 926.
38. Vick, *Illustrated Catalogue* (1865).
39. Beadle, *Canadian Garden*, 346–47.
40. Vick, *Floral Guide* (1874, no. 1), 33–36.
41. Roy Genders, *Collecting Antique Plants* (London, 1971), 238–39.
42. Ibid.
43. Ibid.
44. Ibid., 247.
45. Vick, *Illustrated Catalogue* (1865).
46. M. B. Crane and W. J. C. Lawrence, *The Genetics of Garden Plants* (London, 1939), 43.
47. Charles H. Curtis, *Sweet Peas and Their Cultivation* (London, 1908), 16.
48. Ibid.
49. Peter Henderson & Co., *Seed Catalogue* (1874).
50. Vick, *Floral Guide* (1873).
51. Fraser, *Johnson's Dictionary*, 921.
52. Emsweller et al., "Improvement of Flowers," 905.
53. [Thomas Meehan], "Hints for January," *The Gardener's Monthly* (January 1861), 65.
54. [Thomas Meehan], in *The Gardener's Monthly* (September 1861), 282.
55. Ibid.
56. [Charles M. Hovey], "Floricultural Notes," *The Magazine of Horticulture* (January 1864), 10.
57. Ibid.
58. [Thomas Meehan], "Hints for January," *The Gardener's Monthly* (January 1872), 29.
59. Fraser, *Johnson's Dictionary*, 698.
60. [Charles M. Hovey], "Floricultural Notes," *The Magazine of Horticulture* (September 1864), 220.
61. Joseph Breck & Sons, *Catalogue of Vegetable and Flower Seeds and Bulbous Roots* (1875), 66.
62. Paxton, *Dictionary*, 458.

63. Bailey, *Survival*, 465–72.
64. Ibid.
65. Breck, *Catalogue* (1866).
66. [Charles M. Hovey], "Abronias," *The Magazine of Horticulture* (February 1868), 42.
67. Vick, *Floral Guide* (1875).
68. [Thomas Meehan], "Double Clarkia at Recent London Shows," *The Gardener's Monthly* (September 1861), 283.
69. Vick, *Floral Guide*, (1875).
70. Ibid.
71. Liberty Hyde Bailey, *Manual of Cultivated Plants*, 16th ed. (New York, 1977), 900.
72. Vick, *Floral Guide* (1875).
73. *Hortus Third*, 492.
74. Leroy Abrams, *Illustrated Flora of the Pacific States*, (Stanford, 1960), 4:206.
75. Vick, *Floral Guide* (1875).
76. Harriet L. Keeler, *Our Garden Flowers* (New York, 1910), 200.
77. Abrams, *Flora*, 2:227.
78. Ibid.
79. Jeanne Goode, "Thomas Drummond: A European Naturalist on the American Frontier," *Horticulture* (November 1980), 23.
80. Beadle, *Canadian Garden*, 348.

CHAPTER IV

1. J. S. Ingram, *The Centennial Exposition* (Philadelphia, 1876), 40.
2. Fairmount Park Commissioners, *Catalogue of Tender Plants Grown at Horticultural Hall* (Philadelphia, 1906).
3. *The Centennial Record* (June 1876), 3–4.
4. *Official Catalogue of the United States International Exhibition* (Cambridge, Massachusetts: John R. Nagel and Co., 1876).
5. Thompson Westcott, *Centennial Portfolio: A Souvenir of the International Exhibition at Philadelphia, Comprising Lithographic Views of Fifty of its Principal Buildings, with Letter-press Description* (Philadelphia: Thomas Hunt, 1876), 3.
6. *The Centennial Record* (March 1876), 3.
7. *The Centennial Record* (October 1876), 7.
8. James Buckler, "Horticultural Hall," in *1876: A Centennial Exhibition*, Robert C. Post, ed. (Washington, D.C.: Smithsonian Museum of History and Technology, 1976), 70.
9. *The Centennial Record* (March 1876), 3.
10. World's Columbian Exposition, *Official Catalogue of Exhibits: De-*

partment of Horticulture Building (Chicago, W. B. Conkey Co., 1893), 7.
11. Ibid., 17, 19, 21, 23.
12. *Garden and Forest*, (July 5, 1893), 289.
13. Advertisement material from the World's Columbian Exposition, in the Joseph Downs Manuscript and Microfilm Collection, Henry Francis DuPont Winterthur Museum Library, Winterthur, Delaware.
14. *Garden and Forest* (October 4, 1893), 419.
15. Ibid.
16. W. Atlee Burpee Co., *Burpee's Garden, Farm & Flower Seeds* (1884).
17. Breck, *Catalogue* (1887), xv.
18. Ibid.
19. Liberty Hyde Bailey, *The Standard Cyclopedia of Horticulture* (New York, 1933), 2454.
20. Burpee, *Seeds* (1890).
21. Keeler, *Our Garden Flowers*, 235–38.
22. Curtis, *Sweet Peas*, 18.
23. Joseph Breck & Sons, *Catalogue of Novelties and Specialties* (1887), xiv.

CHAPTER V

1. Jekyll, *Colour*, 74.
2. [William Robinson], in *The Garden* 39 (March 7, 1891), 223.
3. "The Flower Beautiful," *The House Beautiful* (March 1902), 316.
4. G. W. Kerr, "Growing High Quality China Asters," *The Garden Magazine* (March 1912), 83.
5. Emsweller et al., "Improvement of Flowers," 927.
6. J. M. Thorburn & Co., *Thorburn's Seeds* (1905).
7. Kerr, "China Asters," 85.
8. Emsweller et al., "Improvement of Flowers," 927.
9. Ida D. Bennett, *The Flower Garden* (New York, 1910), 111.
10. David Landreth Seed Co., *Landreth's Seed* (1908), 68.
11. *Landreth's Seed* (1914), 89.
12. *Landreth's Seed* (1902).
13. Gertrude Jekyll, *Annuals and Biennials* (London, 1916), 140.
14. Keeler, *Our Garden Flowers*, 493.
15. Joseph Breck & Sons, *Catalogue of Flower Seeds* (1900), 112.
16. Ibid.
17. Wilder, *Colour*, 297.
18. Ibid., 160–61.
19. Jekyll, *Colour*, 52.

20. Ibid.
21. C. Schmidt, "Annual Phloxes," *Gardener's Chronicle* (January 1898), 6–7.
22. Ibid.
23. *The Garden* 39 (March 28, 1891), 293.
24. Jekyll, *Annuals*, 136.
25. Vick, *Floral Guide* (1908), 66.
26. Wilder, *Colour*, 127.
27. Keeler, *Our Garden Flowers*, 493.
28. Alfred Watkins, "Annual Flowers," *Journal of the Royal Horticultural Society* 34 (July 1908), 184–85.
29. Wilder, *Colour*, 200.
30. Jekyll, *Annuals*, 80.
31. Robinson, *English Flower Garden*, 565.
32. Keeler, *Our Garden Flowers*, 478.
33. Adolph Kruhm, "The Annuals Best for Bedding," *The Garden Magazine* (July 1912), 366.
34. Jekyll, *Annuals*, 80.
35. Peter Henderson, *Henderson's Handbook of Plants*, new ed. (New York, 1890), 5–6.
36. Robinson, *English Flower Garden*, 409.
37. Henderson, *Handbook of Plants*, 97.
38. King, *Well-Considered Garden*, 36.
39. Jekyll, *Annuals*, 87.
40. Ibid.
41. Curtis, *Sweet Peas*, 20.
42. Ibid., 22.
43. Crane, *Genetics*, 46.
44. Curtis, *Sweet Peas*, 26.
45. Crane, *Genetics*, 44.
46. Curtis, *Sweet Peas*, 22–23.
47. Crane, *Genetics*, 44.
48. *Landreth's Seed* (1914), 90.
49. Jekyll, *Annuals*, 56.
50. Wilder, *Colour*, 287.

SELECTED SOURCES OF INFORMATION

I. BOOKS

Abrams, Leroy. *Illustrated Flora of the Pacific States*. 4 vols. Stanford: Stanford University Press, 1950–60.

Bailey, Liberty Hyde. *Cyclopedia of American Horticulture*. 6th ed. 4 vols. New York: Macmillan, 1909.

———. *Garden-Making: Suggestions for the Utilizing of Home Grounds*. New York: Macmillan, 1898.

———. *Manual of Cultivated Plants*. 16th rev. ed. New York: Macmillan, 1977.

———. *The Standard Cyclopedia of American Horticulture*. 6 vols. New York: Macmillan, 1914.

———. *The Survival of the Unlike*. 4th ed. New York: Macmillan, 1901.

Baumgardt, John Philip. "A Matter of Taste: Garden Arts and Vogues." In *America's Garden Legacy: A Taste for Pleasure*, edited by George H. M. Lawrence. Philadelphia: Pennsylvania Horticultural Society, 1978.

Beadle, D. W. *Canadian Fruit, Flower, and Kitchen Garden*. Toronto: James Campbell and Son, 1872.

Bennett, Ida D. *The Flower Garden: A Manual for the Amateur Gardener*. New York: Doubleday, Page and Co., 1910.

Breck, Joseph. *The Flower Garden; or Breck's Book of Flowers*. New York: C. M. Saxton, 1861.

———. *New Book of Flowers*. New York: O. Judd & Co., 1866.

Buckler, James. "Horticultural Hall." In *1876: A Centennial Exhibition*, edited by Robert C. Post. Washington, D.C.: Smithsonian Museum of History and Technology, 1976.

Buist, Robert. *The American Flower Garden Directory*. Philadelphia: n.p., 1841.

Cobbett, William. *The American Gardener*. London: C. Clement, 1821.

Copeland, R. Morris. *Country Life: A Handbook of Agriculture, Horticulture, and Landscape Gardening*. Boston: John P. Jewett and Co., 1859.

Curtis, Charles H. *Annuals, Hardy and Half-Hardy.* Garden Flowers in Color Series, edited by R. Hooper Pearson. New York: Frederick A. Stokes Co., n.d.

———. *Sweet Peas and Their Cultivation for Home and Exhibition.* London: W. H. & L. Collingridge, 1908.

Cuthbertson, William. *Pansies, Violas & Violets.* Garden Flowers in Color Series, edited by R. Hooper Pearson. New York: Frederick A. Stokes Co., 1910.

Davis, Julia F., comp. *Art Out-of-Doors: American Gardens, 1890–1930.* Pittsburgh: Hunt Institute for Botanical Documentation, Carnegie Mellon University, and Winterthur Museum and Gardens, 1979.

Earle, Alice Morse. *Old Time Gardens Newly Set Forth.* New York: Macmillan, 1901.

Egleston, Nathaniel Hillyer. *Villages and Village Life with Hints for Their Improvement.* New York: Harper and Bros., 1878.

Emsweller, S. L., Philip Brierley, D. V. Lumsden, and F. L. Mulford. "Improvement of Flowers by Breeding." In *Yearbook of Agriculture, 1937.* Washington, D.C.: Government Printing Office, 1937.

Fairmount Park Commissioners. *Catalogue of Tender Plants Grown at Horticultural Hall.* Philadelphia: Fairmount Park Commission, 1906.

Favretti, Rudy J., and Joy Putman. *Landscapes and Gardens for Historic Buildings.* Nashville: American Association for State and Local History, 1978.

Genders, Roy. *Collecting Antique Plants: The History and Culture of the Old Florists' Flowers.* London: Pelham Books, 1971.

———. *The Cottage Garden.* London: Pelham Books, 1969.

Gilmour, J. S. L., F. R. Horne, E. L. Little, F. A. Stafleu, and R. H. Richens, eds. *International Code of Nomenclature of Cultivated Plants.* Utrecht: International Bureau for Plant Taxonomy and Nomenclature, 1969.

Gorer, Richard. *The Development of Garden Flowers.* London: Eyre and Spottiswoode, 1970.

Harvey, John. *Early Gardening Catalogues.* London: Phillimore and Co., 1972.

———. *Early Nurserymen.* London: Phillimore and Co., 1974.

Haughton, Claire Shaver. *Green Immigrants.* New York: Harcourt Brace Jovanovich, 1978.

Hawthorn, Leslie R., and Leonard H. Pollard. *Vegetable and Flower Seed Production.* New York: Blakiston Co., 1954.

Hays, Samuel P. *The Response to Industrialism, 1885–1914.* Chicago: University of Chicago Press, 1957.

Henderson, Peter. *Gardening for Pleasure.* 2nd ed. New York: Orange Judd Co., 1891.

———. *Henderson's Handbook of Plants and General Horticulture.* New York: Peter Henderson and Co., 1881.

———. *Henderson's Picturesque Gardens.* New York: Peter Henderson and Co., 1901.

———. *Practical Floriculture.* New and enlarged ed. New York: Orange Judd Co., 1874.

Hollingsworth, Buckner. *Flower Chronicles.* New Brunswick: Rutgers University Press, 1958.

Hortus Third; A Concise Dictionary of Plants Cultivated in the United States and Canada, revised and expanded by the staff of the L. H. Bailey Hortorium. New York: Macmillan, 1976.

Hottes, Alfred C. *A Little Book of Annuals.* New York: A. T. DeLaMare Co., 1925.

Hunn, C. E., and L. H. Bailey. *The Practical Garden Book.* 8th ed. New York: Macmillan, 1913.

Ingram, J. S. *The Centennial Exposition.* Philadelphia: Hubbard Bros., 1876.

Jekyll, Gertrude. *Annuals and Biennials.* London: Country Life, 1916.

———. *Colour in the Flower Garden.* London: Country Life, 1908.

Keeler, Harriet L. *Our Garden Flowers.* New York: Charles Scribner's Sons, 1910.

Kemp, Edward. *How to Lay Out a Garden.* New York: John Wiley and Son, 1889.

King, Louisa Yeomans. *The Well-Considered Garden.* New York: Charles Scribner's Sons, 1915.

Lawrence, George H. M. "The Development of American Horticulture." In his *America's Garden Legacy: A Taste for Pleasure.* Philadelphia: Pennsylvania Horticultural Society, 1978.

Leighton, Ann. *American Gardens in the Eighteenth Century "For Use or for Delight."* Boston: Houghton Mifflin Co., 1976.

Loudon, Jane. *Gardening for Ladies; and Companion to the Flower Garden.* 1st American from 3d London ed. New York: Wiley and Putnam, 1843.

———. *The Ladies' Flower-Garden of Ornamental Annuals.* 2d ed. London: W. S. Orr, 1849.

Loudon, John Claudius. *An Encyclopaedia of Gardening.* 5th ed. London: Longman, 1827.

M'Mahon, Bernard. *The American Gardener's Calendar.* 11th ed., edited by J. Jay Smith. Philadelphia: J. B. Lippincott and Co., 1857; reprinted as *McMahon's American Gardener.* New York: Funk and Wagnalls, 1976.

Massingham, Betty. *Miss Jekyll; Portrait of a Great Artist.* London: Country Life, 1966.

Paxton, Joseph. *Paxton's Botanical Dictionary,* revised and corrected by Samuel Hereman. London: Bradbury, Evans, and Co., 1868.

Pennsylvania Horticultural Society. *From Seed to Flower: Philadelphia 1681–1876*. Philadelphia: Pennsylvania Horticultural Society, 1976.
Roberts, H. F. *Plant Hybridization Before Mendel*. New York: Hafner Publishing Co., 1965.
Robinson, William. *The English Flower Garden*. London: John Murray, 1883; 7th ed., 1901.
Sitwell, Sacheverall. *Old Fashioned Flowers*. London: Country Life, 1948.
Smith, Charles H. J. *Landscape Gardening; or Parks and Pleasure Grounds*. New York: C. M. Saxton and Co., 1856.
Tabor, Grace. *Old Fashioned Gardening*. New York: Robert McBryde Co., 1925.
Taylor, Geoffrey. *The Victorian Flower Garden*. Essex: Anchor Press, 1952.
Tergit, Gabriele. *Flowers Through the Ages*. London: Oswald Wolff, 1961.
Thacker, Christopher. *The History of Gardens*. Berkeley: University of California Press, 1979.
Van Ravenswaay, Charles. *A Nineteenth-Century Garden*. New York: Universe Books, 1977.
Van Rensselaer, Mariana Griswold. *Art Out-of-Doors*. New York: Charles Scribner's Sons, 1893; 2d ed., 1925.
Vries, Hugo de. *Plant-Breeding; Comments on the Experiments of Nilsson and Burbank*. Chicago: Open Court Pub. Co., 1907.
White, William N. *Gardening for the South*. New York: A. O. Moore and Co., 1858.
Whitson, John, Robert John, and Henry Smith Williams, eds. *Luther Burbank: His Methods and Discoveries and their Practical Application*. 12 vols. New York: Luther Burbank Press, 1914.
Wilder, Louise Beebe. *Colour in my Garden*. New York: Doubleday, Page and Co., 1918.
Wirt, Elizabeth W. *Flora's Dictionary*. Baltimore: F. Lucas, Jr., 1833.
Wright, Richardson. *The Story of Gardening*. New York: Dodd, Mead and Co., 1934.

II. PERIODICAL ARTICLES

Bailey, Liberty Hyde. "Use of Wild Flowers in Cultivation." *The House Beautiful*, June 1902: 51.
"The Flower Beautiful." *The House Beautiful*, May 1902: 316.
Goode, Jeanne. "Thomas Drummond: A European Naturalist on the American Frontier." *Horticulture*, November 1980: 18–23.
[Hovey, Charles M.] "Abronias." *The Magazine of Horticulture, Botany, and All Useful Discoveries and Improvements in Rural Affairs*, February 1868: 42.

———. "Floricultural Notes." *The Magazine of Horticulture, Botany, and All Useful Discoveries and Improvements in Rural Affairs*, January 1864: 10.

———. ——— -. *The Magazine of Horticulture, Botany, and All Useful Discoveries and Improvements in Rural Affairs*, September 1864: 220.

———. ———. *The Magazine of Horticulture, Botany, and All Useful Discoveries and Improvements in Rural Affairs*, September 1866: 371.

———. "The Progress of Horticulture." *The Magazine of Horticulture, Botany, and All Useful Discoveries and Improvements in Rural Affairs*, January 1865: 1–3.

Kerr, G.W. "Growing High Quality China Asters." *The Garden Magazine*, March 1912: 83–85.

Kruhm, Adolph. "The Annuals Best for Bedding." *The Garden Magazine*, July 1912: 364–66.

"Letter from Paris Correspondent." *The Gardener's Monthly and Horticultural Advertiser*, July 1861: 223.

Loudon, Jane. "Sowing Annuals." *Ladies' Magazine of Gardening*, January 1842: 106.

Manks, Dorothy S. "How the American Nursery Trade Began." *Brooklyn Botanic Garden Record*, Plants and Gardens Handbook Series, 23, no. 3 (Autumn 1967): 4–11.

[Meehan, Thomas] "Books, Catalogues, &c. Review." *The Gardener's Monthly and Horticultural Advertiser*, June 1872: 181.

———. "Double Clarkia at Recent London Shows." *The Gardener's Monthly and Horticultural Advertiser*, September 1861: 282–83.

———. "The Floral Sabbath." *The Gardener's Monthly and Horticultural Advertiser*, September 1872: 269–70.

———. "Hints for January." *The Gardener's Monthly and Horticultural Advertiser*, January 1861: 65.

———. ———. *The Gardener's Monthly and Horticultural Advertiser*, January 1872: 1–2, 29.

———. "Hints for March." *The Gardener's Monthly and Horticultural Advertiser*, March 1872: 1.

———. "Hints for May." *The Gardener's Monthly and Horticultural Advertiser*, May 1861: 129.

———. "Hints for November." *The Gardener's Monthly and Horticultural Advertiser*, November 1872: 1.

Schmidt, C. "Annual Phloxes." *Gardener's Chronicle*, January 1898; 6–7.

Watkins, Alfred. "Annual Flowers." *Journal of the Royal Horticultural Society* 34 (July 1908): 179–88.

III. SEED CATALOGUES

(Titles may vary from year to year. Initials in parentheses are abbreviations for the catalogue collections where issues were located; for the full names of the institutions, see the list in part IV below.)

Breck, Joseph, & Sons. *Catalogue of Vegetable & Flower Seeds.* Boston: 1845, 1866, 1868, 1870–72, 1874–75, 1878, 1881–84, 1888–89, 1894–97, 1900, 1905, 1911. (BH, NAL, MHS, UD)

Bridgeman, Thomas. *Bridgeman's Horticultural Establishment.* New York: 1878, 1906. (NAL, UD)

Burpee, W. Atlee, Co. *Burpee's Seeds.* Philadelphia: 1883–84, 1887–88, 1890–92, 1897–98, 1911, 1915. (BH, NAL, PHS, UD)

Farquhar & Co., Seedsmen. *Gardening in a Nutshell.* Boston: 1898. (UD)

Harris, Joseph, Co. *Harris' Seeds.* Coldwater, New York: 1880–81, 1884–86, 1888–89, 1891–95, 1898–1901, 1903, 1911, 1916. (UD)

Henderson, Peter, & Co. *Annual Seed Catalogue.* New York: 1872, 1874–82, 1884, 1888, 1893–98, 1900–01, 1904, 1907, 1909, 1911, 1915. (MHS, NYBG, UD)

Landreth, David, & Son. *Landreth's Seed.* Philadelphia: 1832, 1866, 1871–75, 1878–79, 1881, 1884–86, 1889, 1891, 1902, 1907. (PHS, UD)

Thorburn, J. M., & Co. *Thorburn's Seeds.* New York: 1838, 1867, 1876, 1890, 1895, 1905, 1913. (UD)

Vick, James, & Sons. *Vick's Catalogue and Floral Guide.* Rochester, New York: 1865, 1868, 1870–85, 1887–88, 1890, 1893, 1896–97, 1901, 1906, 1908. (NAL, MHS, UD, and Longwood Gardens Library)

Wood, T. W., & Sons. *Flower Seeds.* Richmond, Virginia: 1894, 1912. (NAL, UD)

IV. MANUSCRIPT AND SEED-CATALOGUE COLLECTIONS

Cornell University, Ithaca, New York. L. H. Bailey Hortorium (BH). Catalogue Collection.

Henry Francis DuPont Winterthur Museum Library, Winterthur, Delaware. Joseph Downs Manuscript Collection (advertisement material from Chicago World's Columbian Exposition).

———. Printed Books and Periodicals Collection (Chicago World's Columbian Exposition Official Catalogue).

Massachusetts Horticultural Society, Boston, Massachusetts (MHS). Seed and Nursery Catalogue Collection.

National Agricultural Library (NAL), Beltsville, Maryland. Catalogue Collection.

New York Botanical Garden Library (NYBG), New York City. Catalogue Collection.

Pennsylvania Horticultural Society Library (PHS), Philadelphia, Pennsylvania. Pennsylvania Collection (pre-1900 catalogues; Burnet Landreth, Jr., "Address Before the Poor Richard Club of Philadelphia on the 150th Anniversary of the David Landreth Seed Co.," Bristol, Pennsylvania, 1934).

University of Delaware Library (UD), Newark, Delaware. Horticulture Collection.

ZINNIA. DOUBLE.

INDEX

(Note: Appendices are not included in the Index.)

Abronia sp., 2, 3, 57–58, 97
Aster, China. See *Callistephus chinensis*

Bailey, Liberty Hyde, 11, 21, 30, 75, 78
Balsam. See *Impatiens balsamina*
Beadle, D. W., 17, 65
Bennett, Ida, 82
Blanket flower. See *Gaillardia sp.*
Booth, William, 5
Breck, Joseph, & Sons, 5, 10, 35, 54, 56, 60, 64, 75, 76, 77, 84, 99
Bridgeman, Thomas, 5
Buist, Robert, 6
Burbank, Luther, 11, 92
Burpee, W. Atlee, 6, 72, 74, 75, 77, 84, 99

California poppy. See *Eschscholzia californica*
Callistephus chinensis, 1, 3, 20, 24, 30, 36, 46–47, 62, 66, 77, 79–80, 103
Carter, James, and Company, 38, 51, 76
Celosia sp., 1, 3, 7, 20, 36, 39–40, 75, 93
Centennial Exposition, 3, 25, 67, 68, 70, 72–73
Chinese houses. See *Collinsia sp.*
Clark, Captain William, 58

Clarkia sp., 2, 3, 23, 30, 35, 36, 57, 58–60, 97–98
Cobbett, William, 13, 18
Cockscomb. See *Celosia sp.*
Collins, Zaccheus, 60
Collinsia sp., 2, 3, 23, 36, 57, 60, 97
Columbian Exposition. See World's Columbian Exposition
Curtis, Charles, 81, 82, 94, 99, 101
Curtis's *Botanical Magazine*, 65

Darwin, Charles, 10, 31
Douglas, David, 23, 36
Drummond, Thomas, 65
Drummond's Phlox. See *Phlox drummondii*

Eckford, Henry, 99, 101
Eschscholzia californica, 2, 3, 36, 57, 63–64, 72, 77, 87, 92

Fairmount Park, 25, 68
Flanagan and Nutting, 14, 35, 42
Furber, Robert, 51

Gaillard de Morentonneau, 62
Gaillardia sp., 2, 3, 62, 72, 94
Gerard, John, 14, 37
Gilia sp., 2, 3, 23, 35, 57, 61, 97

Haage & Schmidt Nursery, 53, 88

Henderson, Peter, and Company, 5, 15, 19, 25, 51, 72, 97
Hooker, Sir William J., 65
Hottes, Alfred C., 13
Hovey, Charles, 5, 38, 40, 53, 54, 57

Impatiens balsamina, 1, 3, 18, 24, 36, 42–44, 66, 77, 94

Jekyll, Gertrude, 10, 33, 78, 79, 83, 86, 87, 90, 93, 97, 98, 103

Keeler, Harriet, 83, 91, 94
King, Louisa Yeomans, 31, 91, 97

Landreth, David, and Company, 5–6, 15, 19, 35, 72, 77, 82, 84, 102
Lathyrus odoratus, 2, 3, 17, 18, 28, 30, 50–51, 66, 76, 77, 99–103
Lewis and Clark Expedition, 16, 58
Loudon, Jane, 10, 23, 61
Loudon, John Claudius, 21
Lucas, William, 14

M'Mahon, Bernard, 5, 15–16, 17, 18, 19, 38, 42, 46
Marigold. See *Tagetes sp.*
Meehan, Thomas, 9, 10, 23, 25, 27, 53, 78
Mendel, Gregor, 11
Mignonette. See *Reseda odorata*
Miller, Philip, 45
Morse, C. C., and Company, 99

Nasturtium. See *Tropaeolum sp.*

Pansy. See *Viola* × *wittrockiana* (*V. tricolor*)
Parkinson, John, 37, 41

Paxton, Sir Joseph, 39, 54
Petunia × *hybrida*, 2, 3, 7, 25, 30, 55–56, 77, 85–86
Phlox drummondii, 2, 3, 20, 30, 65–66, 88–90

Reseda odorata, 1, 3, 20, 28, 30, 44, 66, 77, 82–83
Robinson, William, 27, 28, 30, 33, 77, 79, 80, 85, 88, 94, 97

Sand verbena. See *Abronia sp.*
Sargent, Charles Sprague, 30, 78
Sweet pea. See *Lathyrus odoratus*

Tagetes sp., 1, 3, 28, 41–42, 66, 72, 87, 97
Thorburn, George, 35, 42, 61
Thorburn, Grant, 5
Thorburn, James, and Company, 77, 80, 81, 82, 84
Tropaeolum sp., 1, 3, 18, 20, 24, 37–38, 74, 81–82

Van Rensselaer, Mariana, 23, 25, 30
Vick, James, and Sons, 5, 7, 19, 40, 50, 51, 54, 58, 60, 72, 80
Vilmorin-Andrieux Company, 6, 52, 53, 58, 72
Viola × *wittrockiana* (*V. tricolor*), 2, 3, 7, 14, 47–50, 62, 75, 77, 84

White, William N., 16
Wilder, Louisa Beebe, 30, 31, 33, 86, 87, 91, 103
Wood, T. W., and Sons, 5, 17
World's Columbian Exposition, 3, 67, 68, 70, 72–73

Zinnia sp., 2, 3, 7, 30, 36, 52–53, 66, 77, 87, 90–91